The PowerPC™ Macintosh® Book

The Inside Story on the New RISC-Based Macintosh

Stephan Somogyi

Foreword by Donald A. Norman

Addison-Wesley Publishing Company
Reading, Massachusetts • Menlo Park, California • New York
Don Mills, Ontario • Wokingham, England • Amsterdam
Bonn • Sydney • Singapore • Tokyo • Madrid • San Juan
Paris • Seoul • Milan • Mexico City • Taipei

Many of the designations used by manufacturers and sellers to distinguish their products are claimed as trademarks. Where those designations appear in this book, and Addison-Wesley was aware of a trademark claim, the designations have been printed in initial capital letters or all capital letters.

The authors and publishers have taken care in preparation of this book, but make no expressed or implied warranty of any kind and assume no responsibility for errors or omissions. No liability is assumed for incidental or consequential damages in connection with or arising out of the use of the information or programs contained herein.

Library of Congress Cataloging-in-Publication Data

Somogyi, Stephan.
  The PowerPC Macintosh book / Stephan Somogyi
    p. cm.
  Includes index.
  ISBN 0–201–62650–0
  1. Macintosh (Computer) 2. RISC microprocessors.
II. Title.
QA76.8.M3S665 1994
004.165– –dc20   93-38276
                                                             CIP

Copyright © 1994 by Stephan Somogyi

All rights reserved. No part of this publication may be reproduced, stored in a retrieval system, or transmitted, in any form or by any means, electronic, mechanical, photocopying, recording, or otherwise, without the prior written permission of the publisher. Printed in the United States of America. Published simultaneously in Canada.

Sponsoring Editor: David Clark
Project Manager: Joanne Clapp Fullagar
Production Coordinator: Gail McDonald Jordan
Cover design: Barbara T. Atkinson
Text design: Joyce C. Weston
Set in 11 point Stone Sans by Shepherd, Inc.

1 2 3 4 5 6 7 8 9 -CRW- 9897969594
First printing, August 1994

Addison-Wesley books are available for bulk purchases by corporations, institutions, and other organizations. For more information please contact the Corporate, Government, and Special Sales Department at (800) 238-9682.

"...we shall never surrender."
—Sir Winston Churchill

Contents

Foreword		**xiii**
Introduction		**xix**
Chapter 1	**How We Got Here From There**	**1**
	RISC at Apple	3
	RISC for the Mainstream	10
	Deal of the Century	12
	RISC System Software	21
	Diversification	23
	How We Got Here From There	28
Chapter 2	**Power Macintosh Hardware Overview**	**31**
	The Big Picture	32
	The PowerPC 601	33
	Direct Memory Access	36
	Memory	37
	Video	41
	Storage and SCSI	46
	NuBus	51
	GeoPort	52
	Ports	52
	Sound	53
	The Power Macintosh Upgrade Card	53
	ABS Hardware	55
	Performance	56
Chapter 3	**Power Macintosh Software Overview**	**57**
	Power Macintosh System Software	57
	Emulation	60

	Native PowerPC System Software	61
	Mixed Mode	62
	Native QuickDraw	66
	Native QuickTime	68
	Memory: Modern and Virtual Both	69
	I/O	71
	INITs and Patches	73
	Software on the Power Macintosh	82
Chapter 4	**An Introduction to Microprocessors**	**85**
	Fundamental Microprocessor Concepts	85
	Architecture	94
	Implementation	97
	Caches	97
Chapter 5	**The PowerPC Family**	**103**
	Now We're Playing with POWER	103
	What Makes a PowerPC a PowerPC?	107
	The Abstract PowerPC	109
	The PowerPC 601	112
	The PowerPC 603	120
	The PowerPC 604	127
	The PowerPC 403GA	132
	The PowerPC 620	134
Chapter 6	**Emulators on the Power Macintosh**	**137**
	Emulation Works	138
	The 68LC040 Emulator	138
	SoftWindows	148
	Wabi	153
	The Bottom Line	157
Chapter 7	**Power Macintosh Hardware in Depth**	**159**
	System Hardware	160
	Upgrade Card	178

	AV Card	179
	VRAM Expansion Card	182
Chapter 8	**Power Macintosh Software in Depth**	**185**
	Mixed Mode	185
	Call Chains	188
	Extensions and Fat Patches	189
	The Code Fragment Manager	192
	Traps	195
Chapter 9	**Looking Ahead**	**199**
	Hardware	200
	The PowerPC Reference Platform	206
	Graphing Calculator	209
	System Software	210
	The Future	214
Appendix A	**Resources**	**217**

Thanks

It is impossible for me to imagine what this book would be like without the extraordinary help of the following people. To claim that they have my heartfelt gratitude would be an understatement.

Jack McHenry, for helping in more ways than I can enumerate and for trusting me enough to let me camp out.

Tim Olson, for pitching in unhesitatingly when things were looking bleak and for providing the core content for Chapter 4.

Jim Gable, for taking the time to read the whole manuscript despite having more than enough other things to do, for the Sunday—night—at—10 phone conversations, and for being a staunch supporter throughout this endeavor.

Maggie Canon, *MacUser's* Editor-in-Chief, for giving me the freedom to get this book done.

Richard Zulch, co-conspirator, for being a high-fidelity sounding board.

Rik Myslewski, fellow raw fish addict of eastern European descent, for deconstruction services and reality checks.

Pam Pfiffner, my fearless leader at *MacUser*, for being quite the understanding boss.

Frank Casanova, for offering help when I needed it.

Richard Clark, for helping out despite maximum entropy in his own world.

More Thanks

In the past, when I've read books that started with "Many people contributed to this book, without whom it wouldn't be nearly as good as it is," or something along those lines, I've dismissed such descriptions, as well as the number of people listed, as exaggerated.

Having finished this, my first book, I can say with some authority that they weren't kidding after all.

The people listed below contributed to this book in one way or another, but not by just "doing their job." They went out of their way to provide information, access, time out of their busy schedules, and other valuable resources when I needed them.

First of all, thanks to those who, among other things, took the time to read parts of this book, often in rather rough form, and provided much-needed feedback despite deadlines of their own: Keith Cox, Michael Dhuey, Bob Hollyer, Gary Kacmarcik, Alan Lillich, Zenon Kuc, Phil Koch, Paul Nixon, and Eric Traut. Any remaining errors are my fault, not theirs.

Thanks also to: Joseph Aseo, Ron Avitzur, Sam Barone, Sheila Brady, Pierre Césarini, Gary Davidian, Ross Ely, Jon Fitch, Don Fotsch, Bill Goins, Carl Hewitt, Ray Jaafari, Annette Machado, Bob Mansfield, Hugh Martin, Jordan Mattson, Brian Mellea, John Mitchell, Dean Mosley, John Nelson, Rolly Reed, Pete Richardson, Greg Robbins, John Sell, Betty Taylor, George Towner, Keri Walker, Gayle Ryan Westbrook, Jim Venable, Paul Wolf, and Mike Yamamura.

Special thanks go to John Hennessey and Dave Patterson, as well as the kind folks at Morgan Kaufman Publishers, who graciously allowed me to use the figure on page 99 from their book, *Computer Architecture: A Quantitive Approach*, which is listed in Appendix A.

My thanks go to Don Norman for agreeing to write the

foreword. His perspective is one that I wish more people in the high-technology industries would adopt.

Thanks also go to my friends in close proximity, who kept me reasonably sane while this thing took over my life for far too long: Nico Kamp & Katy McNamara, Mark Frost, Linda Pitcher, Levi Thomas & Larry Yaeger, and Mitch Ratcliffe.

I'd also like to thank the following people who had no direct input on the content of this book but provided greatly appreciated moral support by checking for life signs and making sure I hadn't imploded: David Biedny, Trudy Edelson, Devon Hubbard, Tom Nielsen, Leonard Rosenthol, and Rich Siegel.

Last but not least, thanks to Carole McClendon for handling the dickering early on, and to David Clark, Joanne Clapp Fullagar, Keith Wollman, and Steve Stansel at Addison-Wesley who thought this was a worthwhile undertaking, and to Gail McDonald Jordan for making it all come together in the end.

Foreword

At first glance, I might seem to be a strange person to be writing the foreword to a book about the PowerPC Macintosh. Until 1993, I was a professor of cognitive science at the University of California, San Diego, a scientist who studied the human mind. Now that I'm at Apple, I serve several roles. As an Apple Fellow, I wander across company divisions as champion of the user. In AppleSoft, our software division, I operate under the title User Experience Architect. Neither of these roles would appear to have anything to do with a new piece of hardware, especially a CPU chip. What has this got to do with either human interface or user experience?

The exciting way to view the new chip is as an enabler for entirely new things that simply could not be imagined before. The PowerPC represents a completely new philosophy and style of CPU for personal computers, which provides entry to a whole new level of affordable performance. This powerful chip will allow users to discover totally new ways of working with machines.

Look, I don't believe the average citizen cares about the technical aspects of computing, such as:

- The operating system;

- The kind of chip used for their CPU;

- How much memory they have;

- CPU speed.

There are several ways to view the power of a CPU based on a reduced instruction set computer (RISC) architecture. One way is simply to look at the speed of the chip: It is incredibly fast. In that sense, the PowerPC is a supercharger. It makes the things we are already doing with computers go faster. That's neat, but it isn't the sort of thing that makes most people's hearts beat faster. Will the average person appreciate that a word processor or spreadsheet is faster? I doubt it.

Many people are led to believe they should care about the technology, but that is only because of the way computers are currently marketed. Computer journalists, especially those who write for the trade magazines, tend to be champions of technology. They, and the salespeople in computer stores, emphasize the technology. But the average person doesn't really care about the details of technology. What we, the everyday users, really care about is getting on with our lives—enjoying our lives. Even the focus on making computers friendly is wrong because it still emphasizes the computer itself.

I care about getting something done: reading the latest news story; seeing the demo my colleague in Tokyo just filmed; learning how sales are doing with our new catalog services; making a reservation at that new restaurant (specifying no smoking, and maybe even peeking at the menu). I care about doing these things and preparing my material so that others can use it, but not about using a computer.

Until now, we have designed machines from the machine's point of view. Computers use information. Invisible. Arbitrary. Difficult. To work, they require precise syntax, details, logic— just the sort of things we are bad at. But there is a mismatch: People are perceptual devices, machines are symbolic.

If we want machines that are easy to use and comfortable for people, we have to make them match people's capabilities. We have to provide perceptual information and minimize the requirements for precise, numerical, or syntactically correct inputs. A graphical user interface such as the Macintosh desktop takes the first steps toward the solution by making heavy use of graphics and menus. But these are primitive steps. The desktop isn't really a desktop, and this graphical user interface isn't really very graphical. The visual appearance is rather flat, more like line drawings and illustrations than rich, visual representations. Up to now, we have lacked the computer power necessary to do more. If we want machines to match people, we need to match the computing power of the brain, or at least of the eyes and ears. Then we can use more natural modes of interaction than keyboard and mouse and arbitrary commands. We can use speech, handwriting, gestures, and whatever else our creativity offers.

The brain is an incredibly powerful device, but it works very differently from our computers. Each element in the brain— the individual neuron—is fairly slow, noisy, and unreliable. It is a semidigital, semianalog device, capable of doing complex signal processing. But there are some 10^{12} neurons, each making an average of 10^4 connections, so interconnected that the apparent slowness and lack of reliability of the individual neuron yields a fast, powerful, robust system. Each of the 10^{16} connections transmits 10 to 100 impulses per second, for a total bandwidth of 10^{17} to 10^{18} impulses per second. The eyes alone generate about 200 megabits of data per second. The brain is a vast, parallel, neural computer that has very different properties from our serial, digital machines.

Computers are good at the stuff we find hard, and bad at the stuff we find easy, such as seeing and walking and talking and—well, all the stuff we all do so well that we take it for granted.

What are people good at? Creativity, humor, emotions, enjoyment. Sports, music, art. What are we bad at? Remembering details, systematic logic, arithmetic, spelling. What are computers good at? Details, systematic logic, arithmetic.

Now that we are moving to much more powerful CPUs, we can begin to make computers that interact with people on human terms. There are several ways in which this new power might be used. Let me point out some that might not be obvious. Consider the conceptual model of a typical application. People are very good at understanding sensible, coherent structures and not so good at understanding or remembering arbitrary commands and actions. This is one of the powers of the graphical user interface over the command-line interface. The real trick in making computers understandable is to provide a coherent, intelligible conceptualization to the users, making sure that all operations and results conform. Today's graphical interfaces do not present a coherent conceptual model. The user may have no understanding of how or why operations get performed. In a spreadsheet, it is difficult to tell the ranges of the functions—just which cells are included in the operation. As a result, when you are using someone else's spreadsheet, it is often difficult to tell just what computations are being performed, and what values are relevant. In a database, it isn't always clear which individual records have been linked, or which operations any query might have to use. Relational databases are often difficult for people to set up and query, in part because they lack an intelligible conceptual model of the operations.

Now that we have the appropriate computer power, we could provide powerful clues to the underlying conceptual model through graphics. Imagine a database query that showed a pictorial rendition of the records and illustrated how a query traversed them, putting together the information for its response. A proper illustration would dramatically improve the user's understanding and make the debugging of failed queries or improperly constructed records much simpler.

Consider educational packages that can make much heavier use of simulation, showing in detail the underlying operations. Today, we have many simulation packages, but they mostly concentrate upon the outcome, not upon showing the underlying processes. Suppose we could illustrate the process as well as the outcome?

What about new methods of interaction, more effective modes? Say gesture, or speech, or handwriting? Or what about using three-dimensional graphics, sound, or speech output? For all of these, we require a lot more computing power. We require the PowerPC.

Today, the user does all the work. Do you want to send a file to a colleague over the network? You must know lots of technical details, including the network path and the name of your colleague's machine. You have to make sure the file and computer protections are appropriate. You have to know if your colleague has the correct applications and fonts. Suppose you could delegate all of this: "Send this OpenDoc document and viewer to Helen." Let the machine worry about the details, bothering you only if it isn't sure which Helen you had in mind or if a serious problem arises. This form of interaction—delegation rather than direct manipulation—requires some inference and general-purpose knowledge by the computer agent that is to do the task—more reason why we need the capability of the PowerPC.

The Power Macintosh provides a new hardware platform, but I think of it as a mere beginning, as an enabler. The truth is, I don't know what the future will bring; nobody ever does. The secret is to be able to take advantage of new potentials, to help us move to another, higher level of capability. This is where we stand today. This book sets the stage by giving you the details you need to take that step. The world of computing has had a prodigious set of advances in the previous ten years. We are now beginning an equally marvelous set of changes during the next decade.

Donald A. Norman
Apple Fellow
Apple Computer, Inc.
Cupertino, California

Introduction

A little more than 10 years ago, a group of stubborn people at Apple Computer shipped the Macintosh and introduced the personal-computing world to a slew of neat new stuff that few users at the time knew what to do with. Those few who "got it" were quickly branded zealots, dismissed as a fringe group, and deemed not part of the Serious Business Computing World.

During the intervening 10 years, the Macintosh has become a computer to be reckoned with in the business world, despite still owning less than 15 percent of the personal computer market. Imitation—litigation notwithstanding—is the sincerest form of flattery, and Windows is doing its best to catch up to the Mac operating system. One of the Mac's great hallmarks has been its integration of system software with hardware. The suggestion that Macintosh hardware is a copy-protection device for the Macintosh operating system has a kernel of truth to it: Apple's strategy has limited the Macintosh environment to its own hardware and, by doing so, limited the proliferation of the Mac but also kept a degree of consistency and compatibility between product generations that is unparalleled elsewhere.

The first generation of Power Macs is a bridge between the past and the upcoming decades. This first generation of Macs using the PowerPC 601 chip are real Macs; no compromises were made to provide extra performance at the expense of compatibility. The ongoing survival of the Macintosh, both hardware and software, depends on the success of these first Power Macs.

The first step in the migration allows Power Mac owners to run their existing Macintosh software, based on the Motorola 68000 (68k) chips on the new machines. Given the investment that current Mac users have in their software, without this compatibility, the Power Macs would be non—starters. Fortunately, 68k compatibility isn't an issue, as the Power Macs' 68k emulator has already proven reliable, compatible, and adequately fast for most tasks.

The second part of the switchover from 68k to PowerPC is centered around native software that takes full advantage of the performance that the PowerPC-based systems are capable of.

What's in This Book

This book is structured to be read sequentially, but it can be used as a reference as well. Those reading it from start to finish will find it increases in technical depth as it progresses. It's designed to provide useful information to anyone interested in acquiring more than a superficial understanding of the first generation of Power Macs and the issues that surround them.

Chapter 1 sets the stage for the Power Macs by going over the history and development of the PowerPC alliance, a previously inconceivable coalition of former competitors. Chapters 2 and 3 provide an overview of the Power Macintosh hardware and software, to get you acquainted with the new machines from a big—picture perspective.

Chapter 4 is an introduction to microprocessors. Differentiation of personal computers today has become increasingly complex and subtle. The ability to distinguish disinformation from useful detail requires more than a cursory knowledge of how the chips work. This chapter offers a painless introduction to the key characteristics and differentiators of different microprocessors.

Chapter 5 introduces the known members of the PowerPC family of microprocessors. Even though the 601 is the only one shipping in systems as this book goes to press, much is already known about its siblings' capabilities.

Chapter 6 is first of the three Power Mac in-depth chapters; it explains the different emulators for the Power Mac and how they work. Chapter 7 offers an in-depth look at the Power Mac hardware, and Chapter 8 does the same for Power Mac software.

Finally, Chapter 9 looks into the future at technologies that are relevant to the Power Macs or PowerPC-based personal computers in general. These first machines from Apple are only the beginning, and some of the future is already visible.

Who This Book Is For

This book is not a step-by-step guide to specific migration strategies. Its goal is to provide the necessary information to allow individuals(or organizations) to make educated decisions about when and how to migrate.

Those who wonder whether to switch to PowerPC now or wait for an even faster Power Mac are doomed to indecision. There will always be a newer, better, whizzier Power Mac just around the corner. The best method for deciding when to switch to the Power Macs is straightforward. Determine the amount of productivity gain you can get from a Power Mac now, and base your decision on that. Most computationally intensive software, especially graphics and desktop-publishing packages, are available now in native versions, optimized specifically for the Power Mac. If you spend most of your time waiting for your Macintosh to catch up, a Power Mac will almost certainly alleviate the problem. Those with compute-intensive software that requires a 68k-based Mac but that isn't yet available in a native version may want to hold off for a while. Bear in mind, though, that the future of Macintosh is PowerPC, the 68k-based Macs' days are numbered.

Users of x86-based PCs may also find this book useful in explaining the features of the new Macs. With the Power Macs, the price/performance ratio is in favor of the

Macintosh for the first time. Existing Windows users can even run most of their Windows applications on their Power Macs with the help of Insignia's SoftWindows.

Now What

I hope this book will get you started on your way into the world of Power Macintosh. The PowerPC alliance and the resultant microprocessors are pitted in head-to-head competition with Intel and its highest-end x86 processors. The Power PC offers the first viable alternative to the x86 because of the performance it offers at a comparatively low price. Although x86 PCs have always been able to offer greater performance for equal or less money, this is no longer the case.

In the long run, adoption of the PowerPC by systems vendors also heralds the beginning of operating system (OS) agnosticism. OSs will no longer be tied to the microprocessor families they run on. The Mac OS runs on 68k- and PowerPC-based systems, and Windows NT will be available for PowerPC machines, as will be some incarnations of IBM's OS/2 and AIX. Non-NT Windows is also available, with the help of SoftWindows, on non-x86 platforms.

The notion of functioning industry alliances is also becoming more accepted. The success of the PowerPC troika is a major counterbalance to the cynicism that developed after the vast number of failed collaborative in past years. The fact that three companies as different as Apple, IBM, and Motorola can work together speaks for the adage that where there is a will, there is a way.

So far, 1994 has seen great changes in the computer industry, and they show no sign of abating. Intel's dominance is being contested not only by PowerPC, but also on the x86 side by companies such as AMD and Cyrix, which are gaining influential allies among systems vendors.

The Power Macs are at the forefront of all this change, and Apple for the first time stands a good chance of gaining significant ground against the installed base of Intel-based Windows PCs.

Soon Macintosh systems will be developed that bring the PowerPC's performance across the entire Macintosh product line. With all these PowerPC-based Macs will come new software that takes full advantage of the available horsepower. Sometimes thereafter, we will no doubt wonder how we ever got along without Power Macs.

Stephan Somogyi
San Francisco, California
May 1994
Somogyi@ZIFF.COM

CHAPTER ONE

How We Got Here from There

he Power Macintosh is a major leap forward in the evolution of the Macintosh.

In hindsight, the change in the Macintosh since its introduction in 1984 has been a gradual one. Memory capacity has grown from 128 kilobytes to hundreds of megabytes, storage capacity has gone from a 400-kilobyte floppy to multigigabyte hard drives, and increasingly powerful members of Motorola's 68000 (68k) processor family have become the engines driving an ever-increasing number of Macintosh computers. However, the performance of later 68k chips—the 68020, 68030, and 68040—didn't increase nearly as much as the capabilities of either memory or mass-storage technology.

Putting a completely different processor, a PowerPC chip, at the core of a Macintosh appears at first to be a profound change, one that makes a Mac not quite a Mac anymore. Far from it.

A Power Macintosh is a Macintosh—an extraordinarily fast one. A Power Macintosh can run all your existing Mac applications—with very few exceptions, thanks to sophisticated software technology—and with the System 7 user interface you're accustomed to. Almost any hardware that you can connect to a 68k-based Macintosh works with a

> **Jargon 101**
>
> RISC stands for reduced instruction-set computer. This term describes microprocessors that were developed with a specific design philosophy in mind. The basic idea is that a chip can perform many simple functions in the same time that a CISC chip can perform a single complex function.
>
> RISC's conventional counterpart is *CISC*: complex instruction-set computer. CISC chips distinguish themselves by performing more complex functions, but with a comparatively steep performance penalty compared to RISC. CISC chips generally perform more slowly, are more complicated to design and manufacture, and are consequently more expensive. The latest CISC chips, such as Intel's Pentium, have many RISC-like features but aren't really RISC chips; they must still perform complex functions like their predecessors for compatibility reasons.
>
> The term *architecture* is used to describe the basic design features that different chips of a processor family have in common, such as number of registers, floating-point capabilities, memory management, and the like. PowerPC is an architecture whose family initially contains the 601, 603, 604, and 620 chips.
>
> *POWER*, the name of IBM's RISC architecture that was the basis for PowerPC, is another one of those great computer acronyms that itself contains an acronym: performance optimized with enhanced RISC.
>
> If you want more details about microprocessor basics, the chips and features of the PowerPC family, and other chips that are competing with PowerPC, see Chapters Four and Five.

Power Mac. But that's not all: New programs written specifically for the Power Macs run many times faster than on the highest-end Quadra.

Fortunately for Mac users, the speed offered by the Power Macs doesn't come at the price of incompatibility. Thanks to Apple's emulator, existing Mac software thinks it's running on a 68k processor.

These new Macs are based on a PowerPC chip, a product of the Apple/IBM/Motorola alliance formed in 1991. But the history of the RISC Mac started about three years earlier, when a group of Apple engineers began building a computer system.

RISC at Apple

Apple has been using RISC chips since the late 1980s and shipped its first RISC-based product, the Macintosh Display Card 8•24 GC, in March 1990. This graphics card used an AMD 29000 (29k) processor running at 30 MHz to accelerate graphics operations on the Macintosh. This acceleration was achieved by replacing code that would normally run on the Mac's own 68k processor with code that ran on the much faster 29000. Since only a specific part of the Macintosh operating system—the QuickDraw graphics software—was replaced by faster RISC code and not the entire operating system, the term *toolbox acceleration* was coined. Toolbox acceleration makes only the most computationally intensive parts of the Mac OS (operating system) run faster. Speeding up select parts of the operating system produces a performance increase perceivable throughout the system. The amount of engineering effort involved in converting only certain parts of the OS was also far smaller than the work required to make the whole thing native. Native software is developed specifically with RISC in mind and takes full advantage of the new processor's performance.

The 8•24 GC card was plagued with incompatibilities, even with Apple's own hardware and software, and it was ultimately abandoned. However, the 8•24 GC was a valuable proving ground for some of the technology found in the Power Macs. The idea of selectively converting the most performance-critical parts of the operating system carried forward to the Power Macs' system software. Power Macs use a hybrid of PowerPC code for QuickDraw, parts of QuickTime, and other compute-intensive parts of the OS, and emulated 68k code for those parts of the OS that wouldn't benefit as greatly from being run on the PowerPC chip.

Entire computer systems based on RISC date back to two distinctly separate RISC projects at Apple that started in the late 1980s.

Jaguar

The Jaguar project officially got under way in summer 1989, although it had been in various stages of planning since mid-1988. The goal of the Jaguar project was to create a microcomputer that had more raw compute-horsepower than any other personal computer on the market and that had a truly human interface that, for example, accepted spoken commands.

Jaguar was to take advantage of RISC's horsepower not only to perform more raw computation in less time, but also to redefine the features of a basic personal computer. To this end, the Jaguar group had its own hardware and software teams. The project was independent of any existing Macintosh projects, much the same way the original Macintosh project was separate from any Apple II–related projects.

Apple's fixation on differentiation from the Macintosh came from Jean-Louis Gassée, at the time the president of

---

Jaguar Spin-Offs

Parts of Jaguar have accompanied recent Macintosh releases, even though Jaguar itself never made it to fruition. The following designs originated in the Jaguar project:
- The industrial design introduced with the Centris 610 as well as the Quadra 800
- The Apple Adjustable Keyboard, which can be split down the middle to angle the two halves so that your hands are held at an ergonomically correct angle

The following were all released with the 68040-based Quadra 660AV and Quadra 840AV and were all results of development work for the original Jaguar system:
- Apple's AudioVision monitor, with its integrated high-quality stereo sound and built-in microphone tuned specifically for speech input
- The GeoPort high-speed telecommunications hardware and software modem technology
- PlainTalk speech recognition, also known internally at Apple as Casper

Apple Products. He insisted that the new machine be completely different from any other computer system. One of the original plans was to use Pink, the code name for a new operating system developed internally at Apple, as the standard operating system for this new machine. Pink ultimately became part of the Apple/IBM negotiations, and the project was spun off from Apple and turned into the joint venture Taligent, which is dedicated to developing and marketing the Pink operating system and related technology as stand-alone products.

Jaguar wasn't initially intended to be a high-volume mainstream system. Instead, mass-market RISC systems would follow sometime later. Shortly after Gassée left in early 1990, however, Apple refocused the endeavor to be a mainstream system: The new computer would be a Macintosh.

In late 1989, the Jaguar engineers started to search for a RISC processor. They visited virtually every RISC chip vendor to determine which chip would suit their needs best.

RLC

While work on the Jaguar was already under way, the early work on the machine that would evolve into the Power Macs began. The core engineering team that designed the Power Macs had previously designed the Macintosh IIfx. As the IIfx's development was nearing an end in late 1989, a pivotal get-together happened during a ski trip in Kirkwood, California. During this trip, the Cognac project was born.

The Cognac project was named obliquely after John Hennessy, a Stanford University professor who is a big RISC proponent and a cofounder of MIPS, the maker of the R4000 family of RISC microprocessors. When the IIfx was introduced in March 1990, the 8•24 GC card, Apple's first product built around toolbox acceleration, was introduced simultaneously. Cognac was an idea for a 68020- or 68030-

based Mac that also contained a 29k RISC chip to accelerate time-critical parts of the OS—more than just the QuickDraw acceleration found on the 8•24 GC.

Another part of the Cognac investigation resulted in a 68020 emulator running on a 29k in software. At the time, the emulator was in the proof-of-concept stages, to determine whether it was feasible to emulate a 68020 in software and whether the resulting emulator would provide good enough performance to be acceptable to users.

Ultimately, the Cognac investigation concluded that it wasn't a financially feasible product. There simply wasn't a way to produce a mass-market version of such a hybrid system with two main processors at a sufficiently low price.

In mid-1990, the 88100-based RLC project got under way. RLC was short for RISC LC, referring to the Macintosh LC, in whose flat box the new machine resided, and whose system software the new RISC-based system was to run.

RLC was designed to be inexpensive to implement, quick to market and exclusively RISC-based. Its goal from the beginning was to support the 68k via emulation. RLC was designed to be as compatible as possible with existing Mac hardware—no changes without good reason. Essentially, RLC took a Mac LC and replaced the 68020 processor with an 88100-based CPU and a 68020 emulator.

RLC and its 68020 emulator were up and running in January 1991. It was able to boot with unmodified 68k-based Mac LC ROMs and run System 7. Early versions of the Mixed Mode Manager—the system software that determines whether code is 68k or for the RISC processor and routes it appropriately—were also put into RLC to allow toolbox acceleration.

Mixed mode is a necessity for native and emulated software to work together seamlessly. The Mixed Mode Manager knows which code is 68k-based and needs to be run by the emulator, and which code is native RISC code that can execute directly on the built-in RISC microprocessor. When mixed mode was first conceived, little thought was given to native apps. These machines were expected to run 68k software in emulation, with an accelerated operating system. The emulator was assumed to run fast enough for this to be a realistic way of running 68k software. Another performance-critical part of the Macintosh OS, the Standard Apple Numerics Environment (SANE) was also converted to run on the 88100. SANE, available in the Mac since the beginning, enables applications to perform floating-point calculations even if no floating-point hardware is present. SANE running native on the 88100-based system drastically sped up floating-point performance for those apps that used it for floating-point calculations.

Searching Out RISC

Using a RISC chip for a personal-computer system that wasn't a workstation was considered daring at the time. Analysts vigorously decried RISC as a fad, since it hadn't caught on in the mainstream personal-computer market. It was evident to the Apple engineers, however, that RISC processors had a brighter future than their CISC counterparts, since RISC had much greater potential for performance improvement over time. The workstation market, with Sun Microsystems in the lead, had already discovered that RISC provided much higher computational performance than the more conventional CISC chip designs.

The Jaguar team eventually picked the Motorola 88110 RISC chip. That decision was not exclusively a result of the existing relationship between Apple and Motorola, but largely a technical one. The 88110 is a single-chip implementation of the 88000 RISC architecture that Motorola first showed the world in mid-1988. At the time, the only implementation of the 88000 (88k) architecture consisted of a three-chip set: an 88100 and two 88200s.

Following Sun's Lead

Sun migrated from 68k-based workstations when it introduced systems based on its own SPARC chip, soon to become the most widespread RISC chip in the workstation world. In some ways, Apple is now following that lead. Although Sun made the transition to RISC with an installed base of far fewer 68k-based workstations than the existing number of 68k Macs, it was nonetheless a radical departure at the time. And history supports its decision—Sun's SPARCstations are successful products.

The Other Contenders

The Jaguar team's initial round of investigation into high-performance processor architectures was comprehensive. The team looked at MIPS' R4000, Sun Microsystems' SPARC, Digital's Alpha, AMD's 29000, Advanced RISC Machines' ARM (used in Newtons), AT&T's Hobbit, Hewlett-Packard's PA-RISC, and even Intel's N10, which was later named the i860. At the time, Apple disregarded IBM's POWER architecture, PowerPC's immediate ancestor, because IBM did not appear inclined to make it available to third parties.

The reasons why each of these architectures fell by the wayside were many and varied. Above all, Apple's executives wanted a partnership with a company that had a solid future and sufficient chip manufacturing capability, and whose chip architecture fulfilled Apple's needs for mainstream computer systems. This also meant that, ideally, Apple would get access to a whole processor family whose members could span the range needed to make low-end, high-end, and portable systems rather than just a single class of computer.

Negotiations between Sun and Apple went quite far: Sun was to use the Macintosh interface as the standard user interface for its UNIX, and in return Apple would use chips based on Sun's SPARC architecture at the heart of its RISC systems. Despite strong proponents of this plan within Apple, the negotiations didn't succeed because Apple felt the manufacturing capabilities for SPARC were insufficient for its needs. At the time, Sun had not yet cut its deal with Texas Instruments to manufacture SPARC chips. Additionally, Apple engineers had reservations about some of SPARC's technical features and the limited breadth of the SPARC family.

The MIPS R4000 family was also a strong contender. In this scenario, the Macintosh user interface would be the alternate user interface for ACE, the

> **The Other Contenders (continued)**
>
> Advanced Computing Environment. Apple would then use the R4000 family of chips for its computer systems. ACE, the consortium that included MIPS, Compaq, and Microsoft, intended to define a standard RISC-based hardware and software environment that would become the equivalent of the x86 standard in the DOS and Windows world. Despite the R4000's technical merits, however, Apple and MIPS didn't come to an agreement largely because Microsoft, Apple's primary competitor on the operating-systems side, was a driving force in the ACE alliance. In addition, MIPS' manufacturing volume was insufficient from Apple's perspective. The ACE consortium later collapsed primarily because of power struggles among its members but also because of Intel's successful lobbying to dissuade systems vendors from using RISC instead of Intel's x86 architecture.
>
> Apple eliminated Intel's i860 mainly because it's fiendishly difficult to write software for. The i860 wasn't designed for the mainstream, and Intel wasn't willing to make the necessary modifications to turn it into a usable chip for an Apple computer. This inflexibility doomed further negotiations.

Motorola's 88000 family was interesting to Apple for several reasons. At the time, Compaq was also investigating RISC chips and its engineers liked the 88110. Both Motorola and Apple were trying to convince Compaq that the 88110 was a good choice. Motorola wanted more high-volume customers, and Apple didn't want to be the only one using the chip.

Apple made the 88110 decision in mid-1990 because Apple engineers considered the architecture sound, and Motorola's intentions for further development meshed well with Apple's plans.

The 88110 chip's feature set, in addition to being a single-chip implementation, was driven largely by Apple's requirements for a mass-market RISC chip. Apple's opinion carried a great deal of weight because Apple's purchasing volume would probably eclipse the combined sales volume of several other RISC chip vendors. Although the projected number of

Power Macs sold is small compared to the projected sales figures for i486- and Pentium-based systems, it is huge compared to the sales generated by the primary RISC market until now: workstations.

After picking the 88000 architecture, the Apple engineers built prototype devices. One of the first was the so-called Cub card, an 88100-based NuBus card. The 88110 single-chip implementation of the 88k was not ready yet; the 88100/88200 multichip solution used on the Cub was close enough to the 88110 so that Apple's engineers could begin work on the emulator and other system-related projects. They also had software development tools that worked with the Cub card, so development could begin quickly. The first version of the 88k-based 68k emulator was developed on the Cub card. The Cub card soon led to the RLC.

RISC for the Mainstream

In early 1991, the Jaguar project was disbanded and folded into the existing Macintosh group. At this stage, Apple's RISC efforts were focused on the mainstream; they were not to be high-end, high-performance, high-price computers anymore. The 88110-based successor to Jaguar was built as a Macintosh and code-named Hurricane.

One of the pivotal points in the 88k-based Mac development came at a sales conference in mid-1991, where RLC was demonstrated to a large audience for the first time. Not only did it run with unmodified LC ROMs and LC system software, but one engineer successfully ran an Apple II emulator for the Macintosh, much to the amusement of all present: an emulator running on an emulator.

RLC's immediate successor was born on another ski trip, this time to Banff, Canada, in March 1991. This machine, housed in a IIsi case rather than the LC case, was the first RISC Mac based on the 88110 rather than the 88100/88200 combination, and it was used to continue work on the parts necessary to make the RISC Mac a viable product.

Even after Apple was well into 88110-based development, some people within Apple expressed market-related reservations about the long-term viability of the 88k family. Despite Ford Motor Company's commitment to using an 88k chip in its next-generation engine computer, no major computer manufacturer had chosen the 88k. The sales volume for the 88k family looked too weak, and Apple didn't want to be the only computer maker using the 88k. So Apple went looking for a RISC chip for the second time.

The first time around, Apple hadn't considered IBM's POWER architecture because it thought IBM was unwilling to let other companies use it. This misconception was cleared up during some of the early high-level talks between upper management at Apple and IBM that ultimately set the stage for the alliance. Even once it became an option, POWER still wasn't Apple's favorite, though: The only implementation of POWER at the time was a seven-chip set. Such a conglomeration was far too expensive and unwieldy for Apple's purposes, and it didn't look as if IBM would be able to design a more suitable version of a POWER processor within Apple's time frame.

While the upper corporate echelons at the two companies were talking about collaboration opportunities, the Apple and IBM engineers met for the first time. This meeting happened to be on a Friday, which was dress-down day at the Austin-based IBM Advanced Workstations and Systems Division, home of the POWER architecture; all the IBMers were in jeans. The Apple engineers, anticipating a meeting with a stereotypical bunch of Big Blue people, were all dressed up in suits. Needless to say, this was an unexpected situation for both sides.

A follow-up meeting, which included Motorola representatives, was held the next week. Apple invited Motorola because of the two companies' long relationship, a result of

> **Clothing Cult**
>
> While the talks were still between engineers and the actual agreements were still off in the future, the IBM contingent, having studied Apple's project-clothing cult, presented the second meeting's attendees with sweatshirts. The sweatshirts bore an IBM logo rendered in Apple's six corporate colors, whose *I* had an apple stem and whose *M* was Motorola's logotype. Having blatantly violated all and sundry trademarks and thrown proper IBM decorum out the window, the head of the IBM delegation was anxiety-ridden about the possibility of the box containing the shirts bursting open on the luggage turntable upon their arrival at the San Jose airport.

Apple using chips from the 68k family in Macs. This relationship continued with the collaboration on PowerPC because Apple felt uneasy about committing the company's future to IBM, one of whose divisions was still a direct competitor to Apple's Macintosh business. Apple involved Motorola not only to have a second source for PowerPC chips, but also because Motorola is one of the few chip manufacturers in the world accustomed to producing quantities of chips in the millions.

Deal of the Century

Apple and IBM have traditionally been archenemies, so the world was surprised to hear about the Apple/IBM/Motorola alliance, whose memorandum of intent was publicly announced in July 1991 and whose details were announced later that year. The alliance consisted of five specific parts.

- Apple, IBM, and Motorola would collaboratively design and build a family of RISC chips, known as PowerPC, derived from IBM's POWER RISC architecture.
- The Apple/IBM joint venture Taligent would be formed to develop, market, and sell a new multiplatform object-oriented operating system that was already under development at Apple.

- Kaleida, another joint venture, would create cross-platform multimedia standards and authoring tools.
- PowerOpen, the specification for a hybrid UNIX system much like Apple's A/UX but that runs on PowerPC systems, would be codeveloped at IBM and at Apple—no specific company was created.

  Based on IBM's AIX version of UNIX, the initial implementation of PowerOpen will provide the ability to run Macintosh software on UNIX-based PowerPC machines much like A/UX does today on 68k-based Macs.
- Apple and IBM would cooperate to integrate Macs into IBM's enterprise networking systems.

Although neither Taligent nor Kaleida has shipped a product and a PowerOpen OS isn't available yet on Power Macintosh either, the PowerPC alliance has already produced its first results: The 601 chip was announced in September 1992, and the 603 chip was announced approximately a year later. The 604 was announced in early 1994, and the 620 should be announced before the end of the 1994.

IBM's POWER seven-chip set, known as RIOS—which is the Spanish word for "rivers" and doesn't have any particular code-name significance—was completely unsuitable for high-volume, low-cost products. These days, RIOS is often referred to as Power1 to distinguish it from Power2, a more recent multichip implementation of the POWER architecture. A project known as RSC, for RIOS single-chip, was already in development at IBM when the Apple/IBM/Motorola negotiations began. The RSC's design goal, however, was to create a straightforward implementation of RIOS without significant modifications.

When IBM's and Apple's engineers got together before the alliance was finalized to discuss how to implement POWER in a way that made sense for Apple, they soon realized that they could use POWER as a foundation, reworking its design and improving the architecture. IBM's willingness to turn POWER into PowerPC greatly contributed to the success of the alliance.

In their meetings, engineers from Apple and IBM recognized several shortcomings of the POWER architecture that prevented its low-cost, high-yield implementation for personal computers. Power1 was, after all, originally designed for workstations with less stringent cost constraints. The IBM engineers took the suggested design changes that resulted from these meetings and convinced IBM management that modifications to POWER were required. Management buy-in at IBM was necessary to override dissenting opinions and instances of "not invented here" syndrome in some divisions.

Similar problems were apparent on the Motorola side. Motorola's contribution to the alliance wasn't only in manufacturing and sales. Part of the PowerPC chips' hardware is based on designs that originated in Motorola's 88110 project. Although it was initially reluctant to share its technology, Motorola's technical contribution to the effort significantly enhanced the PowerPC's value to Apple by minimizing the reengineering of its 88k-based systems.

The successful evolution from POWER to the PowerPC architecture, which made high-speed yet inexpensive single-chip PowerPC implementations possible, is a testament to the willingness of the three companies to overcome significant hurdles in the interest of a mutually beneficial goal.

The first PowerPC chip, the 601, is an amalgam of RSC with enhancements, plus some features of the Motorola 88110. The PowerPC architecture was designed to be more suitable for typical personal-computer tasks, as well as to make evolution and expansion easier. The strengths of PowerPC's ancestors complemented the new architecture: POWER was originally designed for high-performance workstations, and the 88110 had a well-designed interface between the chip and the rest of the computer system. The latter helped conserve the investment that Apple already had in its 88110-based designs. Only minimal hardware modifications had to be made to existing prototype systems at Apple to accommodate the differences between the 601 and the 88110.

Meetings, Phone Calls, Meetings

IBM entered negotiations with Apple because it was interested in having Apple adopt its RISC architecture; IBM also wanted access to the Pink operating system to run on its own hardware platforms. IBM's AIX group in Austin, however, wasn't particularly gung ho about putting the Mac user interface and application services on top of its UNIX. Hammering out the agreements between Apple, IBM, and Motorola took more than six months of frequent conference calls and face-to-face meetings held under the utmost secrecy and often in hastily furnished and otherwise unused IBM and Apple office space. Conference calls twice weekly between Apple and IBM kept everyone up-to-date. These calls were also intended to identify issues that could hinder the deal's completion.

Aside from those negotiating at the executive level, most IBM representatives at the meetings were from the Advanced Workstations and Systems Division, today known as the RS/6000 Division. No one from IBM's Personal Computer division, based in Boca Raton, Florida, was involved at any point. The PC division didn't originally commit to using PowerPC chips and even reserved the right to build MIPS-based systems in case the MIPS architecture turned out to be the RISC standard for Windows. IBM has since created a separate business unit as part of the Personal Systems Division that will build PowerPC-based, nonworkstation computers.

Several landmark meetings with Apple and IBM executives got the entire process under way. One such meeting made RSC+, the initial version of the 601, possible by IBM's agreement to change POWER into PowerPC. Another meeting paved the way for IBM's manufacturing specialists to agree on an aggressive delivery schedule for the 601's chip-production process. The 601 went from design to production in less than two years—a considerable feat.

One of the first weeklong meetings was held in June 1991 in an otherwise abandoned building on IBM's Austin, Texas, campus. The day before this meeting began, IBM installed a network, computers, phones, copiers, furniture, and all the other necessary accoutrements of a well-equipped meeting place. The meeting focused on PowerPC, Pink, and PowerOpen; Kaleida didn't enter the equation until later.

The same group of people reconvened one week later in the opulent Management Development Center training facility at IBM's corporate headquarters in Armonk, New York. Two further weeklong meetings were held here, focused on the same three issues. The final alliance agreement was made at meetings held in Apple's River Park facility in San Jose, California—office space that Apple had recently vacated but whose rent was still paid. Similarly to the first weeklong meeting in Austin, this venue was transformed from an abandoned space to a functional conference facility in a day. This final get-together, during which most everyone lived at River Park, resulted in the full set of Apple/IBM/Motorola agreements, including the birth of Kaleida, that formed what many people within the three companies call the deal of the century.

During the meeting at the River Park facility, each company had areas that were further subdivided by project: multimedia, PowerPC, Pink, PowerOpen, and Networking &

Communication. Because of the large number of attendees, everyone wore stickers, color-coded by company. Apple was red, IBM wore blue, and Motorola green. One of the Apple group's running jokes was that the meeting resembled *Star Trek*: Apple was the Federation; IBM was the Klingon empire with whom they were making peace; and Motorola was the Romulan empire—not because they were the badguys, but simply because they were the makers of the ROMs. Once again, pop culture and high technology collided.

Somerset

The PowerPC part of the Apple/IBM/Motorola alliance came together at Somerset, the PowerPC design facility opened in May 1992 in Austin, Texas. Initially dubbed the Customer Design Center, it was aptly renamed Somerset after the English legend of warring factions in the time of King Arthur laying down their arms and joining forces.

The first PowerPC chip, the 601, was designed primarily at an IBM facility in Austin, with the help of Apple and Motorola engineers, and completed at Somerset. The other members of the PowerPC family—the 603, the 604, and the 620—originated at Somerset. The Somerset facility is singular not just because of its joint-venture nature, but also because of its large staff and because it has the funding to allow the parallel development of multiple chips. The 603, 604, and 620 development went on simultaneously. (Many chip families are still designed in sequence, but even Intel is working on two generations of its x86 processor, the P6 and P7, simultaneously.) As a result of the parallel work at Somerset, all of the initially announced PowerPC chips should have reached first silicon by the end of 1994.

The 601 was produced in record time: a total of 21 months from concept to high-volume production. Development started in October 1991, first silicon happened in September 1992, and volume production started in July 1993. As if this weren't challenge enough, the PowerPC architecture specification was being designed simultaneously with the 601 design, a process that took a total of six months, making life interesting for the chip designers by requiring changes in the chip's design at several steps during development because of changes in the architecture design. The complete PowerPC specification was finished well after work was under way on the 601. Considering that the members of the development team began as strangers, the successful completion of the 601 in that short period of time was no mean feat.

Of all the engineers at Somerset, the Apple contingent numbers fewer than 10. In addition to being vastly outnumbered, they all have dual roles to fulfill: They are chip-design engineers as well as customer representatives to Somerset. Since Apple will almost certainly be the largest single customer of PowerPC chips in the near term, Apple's needs carry significant weight in making design decisions.

The composition of all design teams at Somerset is strictly half IBM and half Motorola, except, of course, if an Apple engineer is part of a team. In the spirit of "trust but verify," this 1:1 ratio ensures that no one company's interest is better represented than that of another. Consequently, no features particularly beneficial to one of the companies make it in to a design. Since the distribution of the companies' employees has been a constant from the outset, much potential disagreement has been avoided outright. Many believe that this is a major factor in Somerset's success. Certainly, no one would have forecast that the 601 would be completed slightly ahead of schedule, especially since the outside world was waiting to hear stories about infighting between the factions.

Mixing Corporate Cultures

Just because the majority of the Somerset engineers are from two companies not known for their casual corporate atmosphere doesn't mean that Somerset is a stuffy place to work. Overall, the Somerset environment is most similar to Apple's. When IBM was interviewing for positions at Somerset, dress-down day on Friday was touted as a perk. This notion is ludicrous to Apple people, and it was soon explained to the interviewers that they were more likely to scare away potential employees than lure them by offering casual Fridays.

Another example of culture clash is the question of drug testing: Apple doesn't, Motorola tests employees randomly with advance notice, and IBM makes testing compulsory. The policy adopted at Somerset was that employees are subject to the personnel policies of their employers; no grand unified policy was set for Somerset as a whole. Another issue was alcohol. No alcoholic beverage is allowed in any IBM building; this is in stark contrast to Apple's traditional Friday-afternoon beer bashes in Cupertino. At Somerset, no alcohol is the norm, since both IBM and Motorola have similar policies.

Even if the predominant mode of operation at Somerset is casual, there are also extremes. In the early days, one member of the Apple contingent always arrived at meetings with a bag of rubber toys. At the start of every meeting, the bag's contents were dumped onto the meeting-room table, and anyone who needed to use a rubber fish to accentuate a point during the meeting could do so.

601 Is Greater Than 88110

After all the political hubbub had died down and the engineers got to work, several things needed to happen. The most important task was to get the first PowerPC chip, the 601, specified, designed, and into production. While meetings between Apple, IBM, and Motorola people were going on, the Apple engineers who'd been working on 88110-based systems needed to shift into gear for PowerPC development. But since a real PowerPC chip was still quite a ways off, they initially used IBM's RSC processor, which was similar enough to allow productive work. The first system-on-a-card similar to the Cub card was the RSC-based Smurf card, named after a little blue thing (as opposed to a Big Blue one). Later versions of the Smurf card were 601-based. The 68k emulator was moved over to it, as were Apple's other RISC software projects.

Because of their active involvement with the 601 design, Apple's engineers were prepared when the first 601s appeared in September 1992. It took only a few days to get the 601 working on a Smurf card, and another two days until the emulator was running well enough to bring up the Finder. This impressive achievement was by far the fastest that Apple had managed to get Macintosh up and running on a new CPU. As one might imagine, it's rather more work to go from 68k to PowerPC than from 68020 to 68030 or from 68030 to 68040. Typically, getting a machine based on a new chip to boot all the way to the Finder takes weeks, rather than days.

After the deal of the century, many engineers both within Apple and within Motorola who had spent considerable time on 88110-based projects weren't too thrilled about switching over to the 601. Nonetheless, Apple's 88110 projects moved over to PowerPC. The Hurricane project switched over to the 601 and soon was renamed Tesseract, and RLC and its IIsi-boxed successor changed names to PDM, short for Piltdown Man. The engineers picked this name because of its symbolism: Piltdown Man was the supposed evolutionary missing link between the ape and *Homo sapiens*. In this analogy, the 68k world was the past, and the future was Jaguar's original spec, a non-Macintosh computer with a vast set of sophisticated but easy-to-use features. PDM, with a RISC core but still a Macintosh, was the missing link.

Since the PowerPC 601 has a virtually identical bus to the 88110, the hardware modifications needed to make the switch from RLC were minimal. The PDM's feature set was chosen deliberately to be less ambitious than that of Jaguar.

The goal, as with RLC, was to minimize the risk involved in introducing RISC into the Macintosh world and to support 68k software via an emulator (by now in its third iteration), but with a high-end Mac feature set. At the time, the Cyclone project was under way, which would result in the Quadra 840av and Quadra 660av. PDM's design is deliberately similar to Cyclone's, since it was to offer a high-end, high-performance Macintosh feature set at an affordable price.

RISC System Software

The development of RISC system software at Apple went through many iterations, much like the hardware. Work on many of the fundamentally new parts of System 7.1.2, the version of system software that shipped with the new Power Macs, began with the Cognac project, where the first 68k emulator running on a RISC chip was developed. The Mixed Mode Manager is another addition to the system, and it has also existed in various forms for several years. However, running System 7 on the Power Macs hasn't always been the clear choice, no matter how obvious it seems today.

Jaguar initially was slated to run Pink, Taligent's new object-oriented operating system, but the migration of the early RISC projects into the Macintosh realm changed the plan back to the Macintosh OS, albeit in a different form. Some within Apple wanted the RISC Macs to take on the workstation market as well, and as a result, they wanted a version of UNIX to be the standard operating system for the RISC Macs. For a month in early 1991, the upper echelons at Apple had to be convinced that the Mac OS, and not A/UX, should be the default operating system.

When the decision to go with PowerPC was made and PDM was in its early stages of development, an ambitious plan for system software was formulated. As attendees of the Apple Worldwide Developers' Conference in 1992 were to learn, the planned evolution of the Mac OS was toward using a microkernel operating system. This new OS would provide features such as preemptive multitasking and hardware memory protection in addition to all the standard Macintosh operating-system services.

The amount of work required to make the Mac OS native and to integrate it with the microkernel was ambitious, to say the least. Although it made sense conceptually to have the next major Mac system-software transition happen simultaneously with the introduction of the new PowerPC-based Macs, the decision was made in July 1992 to scale back the initial RISC system-software effort. For this reason, the more conservative system-software specification, which included the emulator, the Mixed Mode Manager, the Code Fragment Manager, and substantial toolbox acceleration, was dubbed V0, since this version was a step before the original goal of a full microkernel OS. The decision to go with V0 had a major benefit: Since compatibility with existing Macintosh hardware and software is a primary goal for the Power Macs, keeping the changes to the operating system to a minimum greatly reduced the potential for incompatibilities.

When V0 was settled upon as the target, a large effort got under way to determine exactly which parts of the operating system should be made native, to maximize the effect of the PowerPC microprocessor. Many months of investigation and data-gathering resulted in a list of most frequently used parts of the operating system. The more often a particular part of the OS was used, the higher priority it received to be made native. In general, the 90/10 rule is in effect here: 10 percent of the code is used 90 percent of the time. So, to maximize the impact of the PowerPC for system software, finding that 10 or so percent was the key.

In the end, all of QuickDraw, the part of the Mac OS that produces graphics, was made native, as well as many other often-used and performance-critical parts of the OS. Because of the Mixed Mode Manager's ability to switch back and forth between native PowerPC code and emulated code, emulated applications get the full advantage of the native parts of the OS.

Diversification

A big milestone for the PDM project happened in October 1992, at the Apple Pacific sales meeting held in Hawaii. The PDM team was flown out to demonstrate the new PowerPC-based system to a large audience for the first time. The hardware and software for the demo were prepared at Apple, but armed with PowerBooks, the engineers zealously continued software development in Hawaii. Unfortunately, by continuing their work, they wound up breaking the system software.

Panic reigned until the hardware and software were on speaking terms and the demo was stable again—just in time, since it was the night before the demo. Unfortunately, the demo system was left on the stage overnight (the demo was to be a part of the next morning's talks) rather than being locked up and, when the members of the PDM team came to check the machine for the last time in the morning, it was dead. Completely. Perhaps a stagehand had bumped the PDM or inadvertently done something else to cause the machine's untimely demise.

Finally, after taking the system apart completely, reseating all the chips on the motherboard, and putting it back together, it worked again. No one understood why, but they weren't about to question it, since the demo was set to begin within 30 minutes. The machine was gingerly snuck onstage behind a curtain, and when the time came, the demo came off perfectly and the audience of Apple salespeople and executives was none the wiser.

This demo wowed the audience, among whom were Michael Spindler and Ian Diery. It reinforced to all present that the PowerPC-based Mac was a viable product.

> **Code Names for the 7100**
>
> In late 1993, Carl Sagan (the astronomer, not the Mac) became upset upon learning that his name was being used to refer to the midrange of the Power Mac line. He (and his lawyers) sent letters of complaint to Apple Computer as well as to the trade journal *MacWEEK*. From the letter, it was clear he believed Apple was planning to use his name as the product name. As a result of the brouhaha, the engineers changed the name to BHA. BHA is purported to stand for butt-head astronomer.
>
> But it didn't stop there. As a result of Dr. Sagan's actions, the code-name change made it into the national print and radio news and became well publicized. In the first week of 1994, MTV called Apple to check out what was going on. It isn't clear whether it was just calling to check facts or whether it was claiming that "Butthead" was its intellectual property; regardless, the machine's code name was changed once and for all. The Power Macintosh 7100 was finally code-named LAW.

Soon after the demo at the sales conference, it became clear that a single RISC-based Mac wasn't going to be able to fill everyone's needs and that a broader product line was needed. In March 1993, the high-performance variant of PDM, code-named Cold Fusion and later known as the Power Macintosh 8100, was started. Three months later, the midrange machine controversially code-named Carl Sagan, and officially named the Power Macintosh 7100, was started as well, rounding out the product line.

In May 1993, the Apple Worldwide Developers' Conference contained a lot of PowerPC-related technical information and whetted developers' appetites for soon-to-be-available high performance at reasonable prices. However, Apple realized that not all developers would be able to have PowerPC-native products ready in time for the machines' release, so they had to rely on its 68k emulation capabilities. In a successful attempt to assuage developers' fears about the emulator's compatibility, Apple set up a room with prototype PowerPC machines and let developers test their software. The success rate was over 90 percent.

To top it all off, at MacHack in June 1993, the annual Macintosh technical conference and impromptu software-writing event, Apple engineers surreptitiously used a prototype PowerPC machine for people to demo their hacks on, without bothering to tell anyone. The assembled group wasn't told until it was all over that all these often nonstandard and otherwise borderline pieces of software ran on the emulator without a hitch. This was the audience to convince about the emulator's stability and compatibility. Running the hacks on a PDM proved the solidity of the emulator to even the most hardened cynics.

In July 1993, a separate PowerPC upgrade project was started with the intent of providing the most inexpensive PowerPC upgrade possible. The Power Macintosh Upgrade Card, code-named STP, took the minimalist approach and provided only the hardware absolutely necessary for an existing 68040-based Mac to become a Power Mac. The STP solution was also ideal for owners of the Quadra 700, 900, and 950, who would not have the opportunity to get a logic-board upgrade like owners of the Macs that share the same boxes with the three Power Macs.

Apple's PowerPC evangelism efforts began to bear fruit in 1993. Approximately two years before the introduction of the Power Macs, the PowerPC evangelists at Apple began canvassing developers to bring their software native as soon as possible. It was clear that key applications needed to be running native on the Power Mac on, or close to, the date of announcement.

Developers were divided into two camps, InsideTrack and FastTrack, to help them bring their apps native quicker. The small number of InsideTrack developers were those who had apps that Apple considered absolutely crucial to a successful launch of the new machines. The InsideTrack developers started work on their software the earliest and fought through numerous changes in the operating system and in the development tools. The bleeding edge best described where they found themselves. Many of these developers used IBM RS/6000 workstations for development, since the Mac-based development tools weren't available early on. The larger group of FastTrack developers, those with key applications that ideally would be done when the Power Macs were introduced as well, got started later using the Mac-based development tools. Many of them managed to get their products ready in time for the Power Mac announcement, despite the later start.

Many milestones were reached during 1993. System software went alpha in June, and into beta in October. As the shipping date for the Power Macs drew nearer, logistical issues about announcement and availability of systems and upgrades became relevant. The original plan was to announce and ship PDM on January 24, 1994, the tenth anniversary of the Mac, and announce the later availability of the other two machines. This idea was soon nixed by Ian Diery, because he wanted to have not only a full product line available at launch, but sufficient inventory to be able to sell machines to people in volume. Some trade magazines reported this schedule change as a slip—far from it, since at the time the engineering schedules didn't change. The only difference was that there was more time to produce more inventory. The Power Macintosh announcement was also the only Macintosh roll-out where upgrades for previous Mac models were to be available the same day as the new systems. This was also a marked departure from previous announcements, where upgrades were available only many months after the systems' introductions.

One of the final decisions to be made about the PowerPC-based Macs was their names. Speculation about their names ran rampant in the trade press and on online services. There was uncertainty whether they would even be called Macs. Within Apple, this question had an obvious answer: Since one of the primary goals for these machines was total Macintosh compatibility, they were definitely going to be called Macs. But what kind of Macs was the big question within Apple. One faction insisted that they be named Quadras, since buyers were familiar with Quadras and this name connoted the high end of the Macintosh line. It was pointed out, however, that the Quadras were named for their 68040 processor, and that no permutation of 601 resulted in the number 4. Additionally, since the 601's main competitor is Intel's Pentium, it wouldn't look good to have a system name that refers to a lesser digit than the Pentium's 5. The name Power Macintosh seems like the obvious choice now.

Time Lines

Summer 1989: Official start of Jaguar project

Winter 1989/1990: The Kirkwood ski trip and the birth of the Cognac project

March 1990: Release of the Mac IIfx and Macintosh 8•24 GC accelerated graphics card

June 1990: Birth of the RLC project

March 1991: The Banff ski trip and the transition of the RLC project to 88110-based system

June 1991 to September 1991: PDM is born from 88110-based Mac

July 1991: The PowerPC Alliance is announced

May 1992: Somerset opens

September 1992: First silicon of 601

October 1992: Pacific sales meeting in Hawaii

March 1993: Cold Fusion project started

June 1993: Carl Sagan project started

July 1993: STP project started

Late Summer 1993: Decision to provide AV functionality in PowerPC systems at introduction made

January 24, 1994: Original planned date for PDM shipment and announcement of LAW and Cold Fusion

March 14, 1994: Introduction of the Power Macintosh line and its first three members: the 6100/60, the 7100/66, and the 8100/80

How We Got Here from There

To the buying public, new computer systems often seem to appear out of thin air. The evolution of the Power Macintosh line was a long and colorful process. When the first RISC projects started, no one had any idea that the result would be the Power Macintosh. An alliance between Apple and IBM was previously unthinkable, yet this coalition is the basis for Apple's long-term Macintosh plans.

A vast number of people were involved in making these Macs happen—many more than for any other Macintosh, when you count all the IBM and Motorola people involved in the PowerPC effort. What's also amazing about the introduction of the Power Macs is that it happened when Apple originally said it would, in the first half of 1994—even though the prediction was made in October 1991, which, in the computer industry, is a huge time gap.

These Macs are also harbingers of new things to come. PowerPC is the first microprocessor architecture that has any chance of competing with the dominance of x86 machines in the personal-computer world. At the time of writing, the available PowerPC chips provided a vastly better price/performance ratio than competing x86 chips. However, Intel's engineers are good, and the competition for the desktop market will be fierce. It's unlikely that PowerPC will unseat the x86 from its position of dominance anytime soon, but it will almost certainly put a serious dent in the overwhelmingly larger sales volume of x86-based personal computers.

The Power Macs are the first PowerPC-based personal computers to ship, and the future of the Macintosh looks bright. More powerful PowerPC processors are on their way, and improved Macs are already in the works. Many people saw the Macintosh as a doomed system because of poor performance, but rather than achieving parity with the competition, Power Macintosh has allowed the Mac to leapfrog over the high end of the x86 world by providing equal or better performance at lower prices. All these years, Macintosh users have clamored for cheaper yet faster Macs. Here they are.

CHAPTER TWO

Power Macintosh Hardware Overview

he new PowerPC-based Macs don't look all that different from the Quadras. But appearances can be deceiving. The familiar Quadra cases now contain powerful new hardware. But despite their standard Mac features, the Power Macs have the hardware to take advantage of the PowerPC 601's high performance without forgoing compatibility.

This chapter provides an overview of the features of Apple's new PowerPC-based Macs and their differences from and similarities to previous Macs. As you will see, the Power Macs are a combination of old and new. They provide high performance—from 13.5 to over 34 times the performance of a Mac IIci, depending on the task—but are nonetheless priced low. This combination of more compute-horsepower for less money is a result of the Power Macs' hardware design, which has been kept as straightforward as possible by taking advantage of recent Macs' hardware innovations. This design provides most of the features familiar to users of the Quadra 660AV and 840AV, Apple's high-end 68040-based Macs.

The Power Macintosh 6100, Power Macintosh 7100, and Power Macintosh 8100 are the first Macintosh systems based on the PowerPC 601 chip. They are the first stage of Apple's plans to migrate the entire Macintosh product line from Macs based on Motorola's 680x0 (68k) family of microprocessors to PowerPC-based machines. Another part of Apple's migration strategy from 68k to PowerPC is the Power Macintosh Upgrade Card, a deceptively simple card for most 68040-based Macs that turns them into Power Macs without needing to swap out the existing motherboard.

When running native software, which takes maximum advantage of the PowerPC chip's high performance, users can expect to see a 66MHz Power Macintosh 7100/66 perform between two and five times faster, depending on the application, compared to 68k-based software running on a 25MHz 68040-based Quadra 700. Most existing software was developed for the 68k Macs, though; average performance for these applications running on a PowerPC-based Macintosh will be roughly that of the Quadra 700.

But predicting performance of software running on the Power Macs is tricky at best. To get a better idea of how the software side of the Power Macs looks and how to gauge software performance on these new machines, Chapter 3 provides an overview of both system and third-party software for the Power Macs, and Chapter 8 provides an even closer look at how software works on the Power Macs.

The Big Picture

Some of the basic hardware features of the new Power Macs will be familiar to those who have looked at the Quadra 660AV and Quadra 840AV. Although the Power Macs don't have a built-in DSP (digital signal processor) chip, they do use direct memory access (DMA) hardware to move data in the system without burdening the central processor. The Power Macs also sport a 64-bit-wide bus for access to the CPU, RAM, and ROM which allows more data to be moved around at higher speeds than in any previous Macintosh.

They have the same high-speed GeoPort serial ports as well as all the audio features found in the AV Quadras. Users who also want to have the video features offered by an AV Quadra can buy any of the three new Power Macs in a configuration that contains a preinstalled card with all the necessary video-related hardware. Peripherals and NuBus cards that are compatible with AV Quadras and other 68k-based Macs should work unchanged with the three new Macs.

The Power Macs have a standard set of input/output (I/O) ports to connect them to the outside world: ADB, two GeoPorts, audio in and out, SCSI, Ethernet, and the new AudioVision connector for video are available on every Power Mac. The Power Macs' internals also provide a standard Macintosh set of features, but implemented with a clear focus on performance. The hardware in the AV Quadras, the most advanced 68k-based Macs, was the foundation for the Power Macs. But the Power Macs are not engineering workstations. They are designed to be the fastest Macs by far and to provide the highest possible degree of compatibility with their predecessors so that existing investments in hardware and software don't become worthless.

The PowerPC 601

At the heart of each Power Macintosh 6100, 7100, and 8100 is a PowerPC 601 RISC chip, codesigned by Apple, IBM, and Motorola and manufactured by IBM. The 601 chip runs at 60MHz in the Power Macintosh 6100/60, at 66MHz in the 7100/66, and at 80MHz in the 8100/80.

The 601 is the first member of the PowerPC family of RISC chips. Its primary design goals were short time to market and high performance. A 601, which is at the low end of the performance curve for the PowerPC family, performs roughly on par with Intel's Pentium chip—it runs at the same speed—at approximately half Pentium's cost. The

> ### Speed-Bumping
>
> Apple included the clock speed of the 601 in the Power Macs' names so they can perform speed-bumping. As with the Quadra 610, which started as a 20MHz 68040 Centris 610 but was revised to use a 25MHz 68040, speed-bumping increases the frequency of the CPU chip and therefore increases performance. With these new names, Apple can speed-bump a Power Mac without having to come up with a new name while calling attention to the different speed. It's likely that we will see newer, faster revisions of the 6100, 7100, and 8100 that contain 601s running at higher speeds.

100 MHz PowerPC 601, announced in April 1994, is a faster version of the 601. Running at identical speeds, the 601 bests Intel's second-generation Pentiums, running at 90 and 100MHz, in floating-point performance, and it equals their performance when executing typical integer-based code.

One of the 601's particular strengths is exceptionally fast floating-point performance. Floating-point math is used heavily by rendering and animation software. These applications will benefit greatly from going native—that is, adapting 68k-based software to run directly on the PowerPC chip. Floating-point-intensive software has been the exception rather than the rule on the Mac until now. Most applications haven't used floating-point math because it didn't provide enough of a performance boost considering the effort required to integrate it into software. The floating-point performance of the PowerPC family is so great that even developers who wouldn't normally consider using floating-point math are redesigning their software because of the potential speed gain.

System software itself hasn't used floating-point operations for much the same reason as other software. The native version of QuickDraw GX uses some floating-point calculations, however, and therefore benefits a from PowerPC's floating-point capabilities.

Another way of looking at PowerPC's floating-point performance is by comparing the capabilities of the Power Macs with those of the Quadra 660AV and 840AV. Both of these AV Macs have a dedicated floating-point digital signal processor (DSP)—AT&T's DSP3210 chip—built in. The DSP chip handles specific processor-intensive tasks such as software modems. The PowerPC 601 can act as the central processor for the Power Macs and at the same time provide the necessary horsepower to run a software modem without needing a dedicated DSP.

In addition to fast floating-point and fast integer performance, the 601 has other features that make it go fast. The 601 has 32 kilobytes of high-speed cache on the chip itself, so it can store the most frequently used data and code on the chip for the fastest possible access. This reduces the need to fetch code and data from memory, speeding up processing a great deal. Another crucial feature of the 601 is that it has a 64-bit data bus that allows it to move 64 bits, or 8 bytes, of data between itself and the outside world—for example, DRAM, ROM, and an external cache in one fell swoop. Internally, the 601 is a 32-bit chip, though; this means that it commonly operates on 32 bits of data during an operation. It can access 4 gigabytes of memory directly. Four gigabytes is not an arbitrary limit. All memory has an address that lets the CPU chip find exactly the data it's looking for. These addresses usable by the 601 are at most 32 bits large, and the largest 32-bit number is 4,294,967,296, or 4 gigabytes. The 601's 64-bit-wide bus works together with the 601's 32 kilobytes of on-board cache to make it easy for the core of the 601 chip to get at the data it needs, and to send data it's done with back to memory with a minimum of hassle.

The 601 has three other immediate relatives: the low-power 603, the 601-successor 604, and the 64-bit 620, each with different strengths and features. If you're interested in more details about the PowerPC architecture and family of chips, see Chapter 5.

Direct Memory Access

One of the hardware features of the new Power Macs that contributes to their high performance is direct memory access, or DMA. DMA is so critical because it provides help to almost all of the different sections of the Power Macintosh hardware. Its influence isn't limited to just one or two parts of the motherboard.

One of the traditional problems with the Macintosh hardware architecture has been that the 68k CPU chip spent much of its time moving data around rather than devoting itself to running users' software. This was a waste of a CPU chip's processing power.

DMA frees the CPU from having to deal with moving data between peripherals and memory. This means that the 601 in a Power Mac can continue doing whatever it's busy with, without interruption, while data is being read in from the SCSI port or data is sent out over Ethernet. Keeping the CPU out of the nitty-gritty data moving leads to a measurable performance improvement for the user as well as higher-speed data transfer between the Power Mac and its peripherals. The Quadra 660AV and 840AV were the first 68k Macs to take advantage of DMA. The Macintosh IIfx supported DMA for SCSI, serial, and floppy I/O but only if you were running A/UX, Apple's version of UNIX. The Power Macs' DMA is different from that of the 660AV and 840AV, but it fulfills the same purpose.

The Power Macs use DMA for SCSI, Ethernet, both serial ports, sound, on-board video, the floppy drive, and NuBus. Delegating the management of these ports to the DMA hardware lets the PowerPC CPU be used for computationally intensive tasks—such as running a software modem, speech recognition, or full-motion video decompression—without causing the machine to grind to a halt.

The other benefit of DMA is that it allows the 64-bit data bus that the 601 is connected to and the 16-bit-wide I/O bus for the various Power Mac ports to be isolated from each other. This means that the 601 is not only freed from moving data back and forth between memory and the ports, but the 601's bus is kept free of this additional data. This leaves the full bandwidth available to the 601 so it can access important resources such as the Power Macs' ROM without being subject to frequent interruptions, as data from the ports needs to travel over the same bus.

Memory

Access to memory, and the speed at which memory is accessed, is crucial to the performance of any computer system. A high-performance CPU like the PowerPC 601 that can process large amounts of data quickly is especially sensitive to the speed at which memory is accessed. As a result, the Power Macs' designers made the path between the 601 and memory as fast as possible.

Memory in the Power Macs can be one of several kinds.

- Dynamic RAM, or DRAM, is the memory commonly referred to when talking about a system's RAM capacity. DRAM is used to store software while it's running as well as the software's data.
- Level 2 cache RAM can be installed in Power Macs. A system's CPU chip—for the Power Macs, the 601—can get at information stored in the cache much faster than data in regular memory.
- Virtual memory (VM) isn't really memory, it just acts like it: VM reserves space on a hard disk to simulate the availability of more RAM. In reality, the Macintosh operating

system's virtual-memory system swaps data between the hard disk and real RAM to create the illusion of more RAM than is available. On the Power Macs, VM has additional benefits, which are discussed in Chapter 3.

- ROM stands for read-only memory. The 4MB of ROM in the Power Mac contain a large part of the system software, including the 68k emulator. This software in the ROM makes the Power Mac hardware a Macintosh rather than just a PowerPC 601-based computer.
- VRAM, or Video RAM, is used specifically to store video data on a video card. None of the Power Macs have VRAM installed on the motherboard, nor can it be added at a later date. The only VRAM expansion possible with the Power Macs is on the optional VRAM video cards that come preinstalled in 7100 or 8100 models.

Dynamic RAM

Each PowerPC-based Macintosh has 8MB of 80-nanosecond (80ns) Dynamic RAM (DRAM) soldered onto the motherboard. Part of this memory, up to 600 kilobytes, is used up by the video subsystem on the motherboard if a display is connected to the motherboard's video connector at startup.

To expand a Power Mac's memory capacity, each Power Mac has SIMM (single inline memory module) sockets. The SIMMs used to increase DRAM in the 6100, 7100, and 8100 are identical to the SIMMs required for Quadra models such as the 610, 650, 660AV, 800, and 840AV. All three Power Macs, however, have a different number of SIMM slots.

The 6100 has two SIMM sockets that can hold up to 64MB when using 32MB SIMMs, making for a maximum capacity of 72MB RAM. The 7100 has four SIMM sockets and can therefore support up to 136MB. Finally, the 8100 has eight sockets; if filled, they can upgrade this Mac to 264MB.

SIMMs must be added a pair at a time because each of

the standard SIMMs used by the Power Macs provides only 32-bit-wide access to its RAM. Since the Power Macs' data bus is 64 bits wide, RAM must always be added in 64-bit-wide increments.

From a user's perspective, RAM in a Power Macintosh is like RAM in any other Mac. There is no need for special high-speed RAM, as with many RISC-based workstations. Normal 80ns RAM is all that's required to keep the Power Macs running quickly. This ability to use standard RAM makes it much easier to inexpensively upgrade the Power Macs' RAM than would be the case for workstations with comparable performance, and it also lowers the cost of a basic Power Mac.

Intuitively, you might think that using faster RAM might improve the Power Macs' performance. You can install faster RAM—for example, 60ns or 70ns RAM—but your Power Mac's performance will not increase as a result of the faster RAM since it is designed with 80ns RAM in mind. Using 80ns RAM isn't a hindrance for the 601, because it doesn't need to read from or write to RAM nearly as often, thanks to the 601's large 32-kilobyte on-chip cache. In addition, since the 601 can read or write up to 64 bytes of memory during one transaction it can still move a large amount of data in a short period of time. To get any significant boost in performance would require significantly faster, and significantly more expensive, DRAM.

Level 2 Cache

The Power Macs all have a built-in socket that is designed to be home to a Level 2 (L2) cache SIMM. *Level 1 cache* refers to any very fast cache memory that is closest to a microprocessor, such as the 32 kilobytes of cache on the PowerPC 601. A Level 2 cache, which is one step further away from the CPU, consists of very high-speed SRAM (Static RAM), which is much faster than traditional DRAM. This SRAM is physically separate from the CPU chip, but it is connected directly to it via the Power Macs' 64-bit bus. The 8100 ships with an L2 cache SIMM preinstalled that has 256 kilobytes of high-speed 14ns RAM. This cache SIMM is available as an option for the other two Power Macs.

The idea behind the L2 cache is to provide a buffer between the processor and comparatively slower DRAM. To the microprocessor, an L2 cache looks like a part of normal RAM: It keeps most recently used data (and code—the L2 cache makes no distinction between the two) around in case the CPU needs it again soon. The larger the L2 cache, the more data can be kept handy and the less often the system needs to read from or write to DRAM. However, the 256 kilobytes of L2 cache provide the best performance boost for the price. Although larger L2 caches would improve a Power Mac's performance further, adding larger L2 caches approaches the point of diminishing returns, since the performance improvement doesn't increase by the same amount as the L2 cache increases.

Anytime the CPU needs data that isn't in its own L1 cache, it looks elsewhere, and if an L2 cache is present and has the requested data, the CPU can get it much faster than if it had to go all the way to the system's DRAM to get it.

Virtual Memory

Virtual memory (VM), originally introduced with System 7 for 68030- and 68040-based Macs, is also supported on the Power Macs. VM is a feature of system software, but it needs hardware to support it. The 601 chip has the necessary MMU (memory-management unit) built in, just like the 68030 and 68040. But VM in the Power Macs isn't just a port from the 68k version; it's been completely revamped and greatly improved for the PowerPC to be faster and to provide additional memory savings for native applications.

Chapter 3 explains the new Power Macs' virtual-memory software in detail.

ROM

Each Power Mac has 4MB of ROM built in. This is twice as large as the previously largest ROM, the 2MB of ROM in the 660AV and 840AV machines. The Power Macs' ROM contains much of the system software for the PowerPC Macs, including the 68k emulator that allows Power Macs to run 68k software as if they also had a 68k chip built-in. One side effect of this large ROM is that the size of the 7.1.2 System file on disk is significantly smaller than that of, say, a 7.1 System file on a Quadra 700. The other benefit is that the OS itself takes up less space in RAM as well, since code can be run directly in ROM, without the need to copy it to RAM first.

Video

All three PowerPC Macs come with built-in video support on the motherboard. The video options available on the new Power Macs have something for everyone:

- The basic built-in video on the 6100 suffices for most standard applications.
- The AV Card provides the video side of Apple's AV technology.
- The VRAM cards for the 7100 and 8100 provide the mainstream yet high-performance video solution.

Unlike the Quadras' built-in video support, which uses separate RAM dedicated to video, the 6100, 7100, and 8100 allocate part of the system's RAM when using motherboard video. If you look in the About This Macintosh window in the Finder, any DRAM allocated to video is part of the memory shown as used by the System and isn't separately identifiable.

Built-in Video

The built-in video hardware in the Power Macs has no more than a passing resemblance to motherboard video on previous Macs. A side effect of this video architecture that uses system RAM rather than dedicated Video RAM is that, depending on the size of the monitor and the desired bit depth, it can use significant amounts of system resources. A Power Macintosh running internal video will slow down perceptibly as bit depth increases. Although the slowdown isn't debilitating, it's noticeable at higher bit depths. If you regularly need to use thousands of colors, you should consider avoiding internal video. A VRAM-based alternative is available for the 7100 and 8100; it is discussed later in this chapter.

The connector for the on-board video on the rear of the machines is Apple's HDI-45 AudioVision connector. Since the only display that directly supports this connector is Apple's own AudioVision display, an adapter cable that converts the AudioVision connector into the standard DB-15 video connector used by most monitors comes with every Power Macintosh 6100. Since the other Power Macs come with additional video interfaces, they are not shipped with this adapter.

If no display is connected to the motherboard's display connector—for example, on a headless server or when you are using another video card—no memory is allocated for on-board video. All the memory is available for the operating system and applications. With a display connected, the built-in video supports the displays at the bit depths shown in Table 2.1.

The worst case is the 13-inch display in 16-bit mode; 600 kilobytes of RAM are needed to support this display mode. However, even if a video mode that uses fewer than 600 kilobytes is used, the Power Macs allocate the full 600; this is necessary if the user ever wants to increase the bit depth;

Cycle-Stealing Video

The Power Macs' on-board video subsystem shares the RAM on the motherboard with the rest of the system. There is no separate Video RAM, as on the Quadras. This has the benefit of being less expensive, since there is no need for specialized RAM, and also less complicated from a systems design perspective. The drawback to cycle-stealing video, though, is performance.

On most personal-computer systems, there's only one path into and out of the system's RAM. If the CPU chip wants access to information in RAM, it must use that path. Most interaction with RAM is relatively short, since moving large quantities of information in and out is rare. Video on the Power Macs is an exception to the rule, since a single screen refresh might need to move up to 600 kilobytes of data. Since a refresh happens roughly 60 to 75 times per second, depending on the monitor being used, in the worst case a little less than 35 percent of the entire system's bus bandwidth is used to transport video data.

Although this may seem like an unreasonable performance penalty to pay, in reality it isn't that bad. The Power Macs' Data Path chips, which are explained in Chapter 7, shield the 601 and all the parts of the system connected directly with the 601 from the constant video data flowing by. This allows the 601 to get on with its business, while the main effect of all the video data traveling across the bus is to limit the amount of available bandwidth. In addition, 98 percent of the 601's memory accesses stay on the chip; only 2 percent actually make it out onto the CPU bus. The only time that on-board video gives the 601 any pause is if it is trying to read from memory while a video refresh is going on. This situation is known as *bus contention*; in such instances, parts of the Power Mac hardware have different priorities that determine their access to the bus. The higher the priority, the less other subsystems in the Power Mac can hog any part of the bus. The video subsystem has the second highest priority, since its refresh must happen without fail at exactly timed intervals.

Users who need higher video performance will have either an AV or a VRAM card. And most users of on-board video will not be using the worst case: 16-bit color on a 13-inch monitor. The most common case will probably be 8-bit color on a 13-inch monitor, which uses around 15 percent of the system's bandwidth—not inconsequential, but by no means debilitating, either.

Table 2.1 Displays Supported by Built-in Video

Display	Resolution	Bit Depths	RAM Used at Max Bit Depth
12" RGB	512 × 384	1, 2, 4, 8, 16	393216 bytes
13" RGB	640 × 480	1, 2, 4, 8, 16	614400 bytes
15" Portrait	640 × 870	1, 2, 4, 8	556800 bytes
16" RGB	832 × 624	1, 2, 4, 8	519168 bytes
VGA	640 × 480	1, 2, 4, 8	307200 bytes

for example, to look at a color photograph. If the whole amount weren't already reserved, the user would either have to restart the machine or be faced with having to look at the image in an undesirable bit depth.

However, internal video isn't the only option for the Power Macs. Apple ships higher-performance video cards with some of them. Both the Power Macintosh 7100 and 8100 come with one of two video cards installed in their processor direct slot (PDS).

Processor Direct Slot

The processor direct slot (PDS) is a direct connection to the motherboard's 64-bit system bus. Any card in this slot has the same access to system resources as if it were directly on the motherboard. This is in contrast to expansion buses such as NuBus, which have significantly lower throughput. NuBus is a 32-bit bus that generally runs at 10MHz; a 6100/60's 64-bit bus runs at 30MHz, half of its 60MHz CPU's speed, but still provides six times the bandwidth of NuBus. An 80MHz 8100/80's bus runs at 40MHz, eight times the throughput of NuBus.

All three Power Macs have one PDS. The 6100's is unused by default but can contain either an AV Card or the Power Macintosh NuBus Adapter Card. The 7100's and

8100's PDSs are always used for one of two video cards: the AV Card or the Power Macintosh VRAM card. The VRAM card is not available as an option for the 6100.

AV Card

The AV (audio/video) versions of all three PowerPC Macs have the AV Card installed, adding another video-output option. The AV Card, which plugs into the processor direct slot (PDS), contains a less expensive but equally capable version of the AV subsystem introduced in the Macintosh Quadra 660AV and 840AV and provides the same AV features as these Macs. The AV Card supports composite video and S-video in and out and has a standard DB-15 video connector. For more details about the AV Card's capabilities, see Chapter 7.

The digital audio video (DAV) connector in the Quadra 660AV and 840AV has changed for the Power Macs, since the AV hardware no longer resides on the system's motherboard. The DAV connector allows direct access to the digital audio and video data coming from the AV hardware. This degree of access is necessary for cards that directly process the audio and video data and need the highest possible performance. NuBus simply doesn't have the bandwidth to support continuous streams of audio and video data.

The DAV connector on the AV Card has the same electrical connections as the original DAV, but it is redesigned for use with a ribbon cable rather than as a plug-in slot for a card. With the new scheme, NuBus cards that have DAV support connect to the AV Card via a ribbon cable. Older cards that use the inline DAV slot on the 840AV (the 660AV doesn't have a DAV slot) are not easily usable in the AV configurations of the new Power Macs, since the location of their DAV connector is on the underside of the card, rather than in a location more accessible to the DAV ribbon connector.

The AV Card comes with 2MB of VRAM installed on the card, and the VRAM capacity of this card can't be expanded.

VRAM Cards

Those 7100s and 8100s that don't come with the AV Card have a Power Macintosh VRAM card installed in the processor direct slot (PDS). As a result, every 7100 and 8100 ships with dual-display support as a standard feature. Out of the box, you can connect your Power Mac to two monitors without any additional video hardware. The VRAM frame-buffer card comes in two variants: one for the 7100 with 1MB of VRAM, expandable to 2MB, and one for the 8100 that comes with 2MB, upgradable to 4MB. The VRAM frame-buffer card can support the displays and bit depths shown in Table 2.2.

QuickDraw runs native on PowerPC Macs. Since QuickDraw isn't emulated, a VRAM card in a Power Macintosh will perform very well, rivaling and often exceeding the performance of many NuBus video cards.

Storage and SCSI

As with all Macs since the Mac Plus, the Power Macs use a SCSI bus to connect to mass-storage devices such as hard drives, CD-ROM drives, DAT drives, removable media, and other devices such as scanners and printers. The SCSI

Table 2.2 Displays Supported by the VRAM Cards

Power Mac Model	MB VRAM	Display Size	Resolution	Maximum Bit Depth
7100/66	1	12"	512×384	16
	1	13"–14"	640×480	16
	1	15"	640×870	8
7100/66 or 8100/80	2	16"	832×624	24
	2	21"	1152×870	16
8100/80	4	21"	1152×870	24

connector on the back of a Power Macintosh is the same 25-pin connector that Mac users have been accustomed to since the introduction of SCSI with the Mac Plus. Any SCSI peripheral that is compatible with the Quadra 660AV or 840AV should work with the new Power Macs.

Like the Quadra 660AV and 840AV, the standard SCSI bus of each Power Mac supports throughput up to 5MB per second. The Power Macintosh 8100 has a second, independent SCSI bus that supports Fast SCSI throughput up to a theoretical

---

Upgrading SCSI

If you're upgrading from a 68k-based Macintosh to a Power Macintosh, you should be aware of a few problems that you may encounter on the path of migration. Owners of the 660AV and 840AV Macs have experienced SCSI-related problems that manifest themselves most commonly as random but frequent crashes, none of which are ultimately the fault of the new Macs. Since the SCSI implementation—both hardware and software—of the Power Macs is so similar to that of the 68k-based AV Macs, users upgrading may experience similar challenges and be tempted to blame them on the new Macs.

The main sources of problems with SCSI on the Mac are traceable to either SCSI cabling or termination. Make sure you have high-quality cables and proper termination. Most SCSI problems that appear to be the fault of new Macs are cabling or termination problems. Recently, vendors have begun offering digital active termination schemes. Such terminators generally improve the signal quality on SCSI chains. Many SCSI problems on the 660AV and 840AV, as well as on the Power Macs, have cleared up as a result of installing this kind of terminator.

The SCSI driver software installed on drives is another facet of SCSI devices that you shouldn't ignore. Apple introduced the new SCSI Manager 4.3 with the Quadra 660AV and 840AV, and this new SCSI system software is also part of the Power Macs. If you want to get the best possible performance from your SCSI peripherals, make sure your drivers all take advantage of SCSI Manager 4.3's features. (Old drivers work fine, but do not offer peak performance.)

If you want more information about things to watch out for when upgrading from your 68k Macs to a Power Macintosh, see Chapter 7.

maximum of 10MB per second, similar to the second internal SCSI bus in the Quadra 900 and 950.

Each of the three new Macs is available with a double-speed internal CD-ROM drive. These are the first drives from Apple that don't require a caddy to use discs. Previously, to use a CD-ROM disc in a drive, you had to place it in a holder called a *caddy,* and then the caddy went into the drive. The drives in the Power Macs behave just like audio CD players: They have a small button on the front that you press to make the CD drawer scoot out. Place the CD in the drawer, press the button, and the drawer scoots back in and brings the CD up on your Mac's desktop.

Double-speed refers to the capability of the drive to spin the CD-ROM at twice the normal rotational speed. Despite their moniker, double-speed drives rarely perform at twice the speed of a conventional CD-ROM drive. The maximum amount of data that a double-speed drive can feed a Mac is around 300 kilobytes per second, quite pokey when compared to hard disks. Higher-end hard drives typically allow transfer rates of around 3.5 to 4MB per second. At this writing, CD-ROM drive manufacturers were introducing triple- and quad-speed CD-ROM drives—great for CD-ROM users, but still slower than a hard disk. To take fullest advantage of a quad-speed drive's ability to transfer data quickly to the Mac, the data on the disc must be arranged just right. Unfortunately, this arrangement can make for much slower reading on non-quad-speed drives.

You can use the internal CD-ROM drive in a Power Macintosh for audio-playback and, with QuickTime's built-in support for the audio-extraction features of these drives, you can also use it to convert audio tracks on an audio CD into QuickTime movies. All AV versions of the three Power Macs come with internal CD-ROM drives preinstalled. Some non-AV versions also come with CD-ROM drives.

Other SCSI devices can be installed inside the new Power Macs. Like the Quadra 610 and 660AV, the Power Macintosh 6100 has two internal bays for SCSI devices. One is a 5.25-inch half-height bay suitable for a CD-ROM drive; the other is a 3.5-inch one-third-height bay for a hard disk. Since all Power Macs come with hard drives, the hard-drive bays all contain a drive. The Power Macintosh 7100 has the same two internal bays as the Quadra 650. And like the Quadra 800 and 840AV, the minitower 8100 has both the internal bays that the other two Power Macs' cases provide, plus a 3.5-inch full-height internal bay for a large-capacity hard drive. The large bay in the 8100 box is intended for high-performance drives, which should be connected to the 8100's internal-only Fast SCSI bus. See Figure 2.1.

The 8100 has a second independent Fast SCSI bus that is accessible only via an internal connector, identical to the internal SCSI ribbon connector used for all Macs. This bus is internal-only for several reasons. To achieve the highest possible throughput, this bus must remain as free as possible from electrical noise. With the wide variety of SCSI devices and cables available, an external bus is rarely clean enough to support such high throughput.

Although active termination and high-grade cables minimize the problems, there are other good reasons to keep the Fast SCSI bus internal. Even most hard drives would have a hard time maxing out the 5MB per second bandwidth of the external bus. Minimizing the number of devices on the internal bus reduces the possibility of multiple devices having to contend for available bandwidth on the Fast SCSI bus. If the Fast SCSI bus had an external connector, everyone would connect their existing SCSI chains to it since, after all, faster is better. All the slow devices would have to share the bandwidth of the Fast SCSI bus, severely reducing its utility for real fast drives. Today's double-speed CD-ROM drives deliver no more than 300 kilobytes per second

FIGURE 2.1
The Power Mac 6100, 7100, and 8100

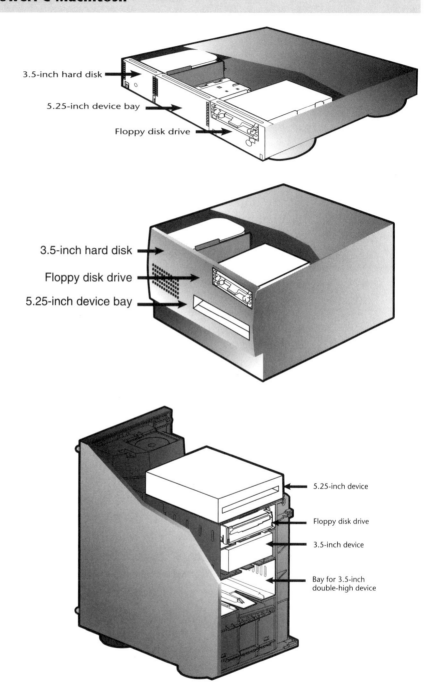

throughput; Apple connects any internal CD-ROM drive in the Power Macintosh 8100 to the standard SCSI bus, leaving the high-speed bus free for devices that can take advantage of its bandwidth.

NuBus

The 7100 and 8100 both have three NuBus slots for expansion, and the 6100 can be outfitted with a NuBus adapter card that plugs into its processor direct slot (PDS). Once installed, this adapter supports one 7-inch NuBus card. Most NuBus cards that work in the 660AV and 840AV should function without problems in the Power Macs.

The Power Macs support NuBus 90, which is a more recent version of the original NuBus 87 specification that has been used in all NuBus-endowed Macs prior to the Quadra 700. NuBus 90 can provide higher performance than NuBus 87 in its *burst* mode, when transferring data from one NuBus card to another.

NuBus, however, may not be as crucial for the new Power Macs as it has been in previous 68k-based machines. Traditional inhabitants of NuBus slots have been video cards, Ethernet cards, and, more recently, audio/video capture cards. Since Ethernet comes on the motherboard of every Power Macintosh and the 7100 and 8100 have either a VRAM card or an AV Card preinstalled, much of the need for NuBus expansion has been eliminated. Nonetheless, these two Macs sport three available slots each.

The NuBus adapter for the 6100 can't be used in the AV versions, since the 6100 motherboard has only one processor direct slot (PDS) that can be occupied either by the AV Card or by a NuBus adapter. Other NuBus adapters, such as the ones for the IIsi and for the Centris/Quadra 610 or 660AV, cannot be used in the 6100. The only specific requirement for a NuBus card to work with 6100's NuBus adapter is that it must conform to the short 7-inch NuBus card-size specification; this requirement is the same for the 610 and 660AV adapters. Cards that work in these machines should work in a 6100.

GeoPort

The GeoPort software for Power Macs runs on the PowerPC chip. No digital signal processor is needed. The GeoPort features aren't tied with the video capabilities of the AV Card, either. Consequently, every Power Macintosh, including those without AV Cards, can run a GeoPort modem.

The GeoPort hardware and software, introduced with the Quadra 660AV and 840AV, used the DSP3210 digital signal processor built into those machines to provide a software modem. All the work that dedicated modem hardware would normally perform was done in software running on the DSP chip in the AV Quadras.

Both serial ports on the Power Macs can be used as GeoPort ports. With the addition of a GeoPort Telecom Adapter connected to a Power Mac's modem port, and with the necessary software installed, even the 601 in the Power Macintosh 6100 has the oomph to also act as a 14400bps V.32bis modem, but it must have the hardware support of DMA to move all the data back and forth in the Power Mac system. Without DMA, a software modem would not be possible.

The slowdown caused by the compute-intensive task of running the modem is perceptible by the user, especially on the slower Power Macintosh configurations. The slowdown is by no means debilitating, however; the Power Mac is still responsive when the GeoPort modem is in use.

Ports

Except for the HDI-45 AudioVision connector, none of the ports on the back of the new Power Macs are unfamiliar to users of earlier Macintosh models. They all have a DB-25

SCSI connector, an AAUI Ethernet port, two GeoPort serial connectors (modem and printer), ADB (for the keyboard and mouse), an audio-in jack, and a speaker jack, just like the 660AV and 840AV. As with these two 68k-based Macs, the sound subsystem in the Power Macs supports high-quality 16-bit stereo sound.

These ports work with the same devices as 68k-based Macs, so buying a Power Macintosh means that you can still use your old peripherals. For 6100 owners, the bundled HDI-45 to DB-15 adapter allows existing monitors to be connected without hassle. The additional microphone necessary for PlainTalk is supplied separately and plugs right into the sound-in jack in the Power Macs.

Sound

All three Power Macs have high-quality sound input and output capabilities built in. None of these features require an AV Card, nor will an AV Card improve a Power Mac's sound features.

Each Power Macintosh has a sound-in and a sound-out jack on its back panel. These stereo jacks are the same size used on portable stereos. The sound hardware inside the Power Macs supports 16-bit digital audio, in and out, at a sample rate of 44.1kHz. This sample rate is the standard rate for audio compact discs.

With these high-quality sound capabilities, the Power Macs support Apple's PlainTalk speech-recognition and speech-synthesis software. This software is available from Apple, together with a microphone designed specifically for speech recognition, as an added-cost option.

The Power Macintosh Upgrade Card

The Power Macintosh Upgrade Card is the most inexpensive way of upgrading a 68040-based Macintosh to a Power

Macintosh, but it doesn't offer the same performance as a Power Macintosh. The Upgrade Card is deceptively simple. It plugs into the 68040 processor direct slot (PDS) in a Quadra 700, 900, 950, 650, or 610 and contains a PowerPC 601 chip, 1MB of 15ns Level 2 cache, and 4MB of ROM. The ROM on the Power Macintosh Upgrade Card is almost identical to that in the 6100/60, 7100/66, and 8100/80, except for minor differences necessary for the card to run in the Quadras. The ROM contains the 68k emulator as well as many other new PowerPC-related system-software features such as Native QuickDraw.

The Upgrade Card is designed so that its 601 always runs twice as fast as the 68040 in the host Mac. The 601 runs at 40MHz in an original Centris 610 whose 68040 runs at 20MHz. In a Quadra 700, whose 68040 runs at 25MHz, the 601 runs at 50MHz. In a 33MHz Quadra 950, the 601 on an Upgrade Card runs at 66MHz. There aren't different versions of the Upgrade Card—one size fits all.

A Quadra with an installed Upgrade Card can boot up in one of two modes:

- The 68040 is active, in which case the user can run 68k software native.
- The card boots up in PowerPC mode to allow native PowerPC apps to run.

When the Upgrade Card is active and the Mac is running in PowerPC mode, it cannot use the 68040 on the Quadra's motherboard to run 68k software. If the 601 is the active CPU, the 68k emulator is used to run 68k software.

This ability to run in either mode is an excellent feature for those who want to make the most painless transition to PowerPC. Only a restart separates users from running native PowerPC apps and running 68k apps at full speed on a 68040.

The performance of an Upgrade Card is less than that of a Power Mac, since Power Mac hardware features such as DMA and a 64-bit bus aren't available on the Macs that accept Upgrade Cards. The 601 alone provides a great deal of performance, and the large 1MB Level 2 cache on the Upgrade Card compensates for this problem. But having to go through the 32-bit 68040 PDS, which requires translation from the 601 bus to and from the 68040 bus, creates a bottleneck. The 68040 bus acts like a funnel and limits the amount of data that can move between the card and the motherboard.

ABS Hardware

Apple Business Systems (ABS) is also moving toward the PowerPC with its servers. The two lower-end offerings are identical to two of the Power Macs, and the high-end PowerPC-based Workgroup Server is a new development with a few significant changes. For more information about the software running on these servers, see Chapter 3.

Apple Workgroup Server 6150

The Apple Workgroup Server (AWS) 6150 is identical to the Power Macintosh 6100 except that it has a 512-kilobyte Level 2 cache preinstalled. The AWS 6150 runs at 60MHz and, except for the different front panel and the L2 cache, it is exactly the same machine.

Apple Business Systems offers upgrades from the AWS 60 to the 6150.

Apple Workgroup Server 8150

The AWS 8150 is identical to the 8100, except for its preinstalled 512-kilobyte Level 2 cache. There are no other differences other than the front panel.

Upgrades are also available to help users migrate from the Apple Workgroup Server 80 to the AWS 8150.

Apple Workgroup Server 9150

The Apple Workgroup Server 9150 has no counterpart in the Power Macintosh line. The 9150 is a further development of the Power Mac 8100 hardware design and includes all of its features, including the two SCSI buses. The 9150 has two differences: It has a fourth NuBus expansion slot as opposed to the 8100's three, and the 9150's motherboard is designed to fit into a Quadra 900/950 box.

This case is the same one used by the Apple Workgroup Server 95; the 9150 is the upgrade for the 68k-based AWS 95. However, the AWS 9150 does not run A/UX and AppleShare Pro like the 95; it runs AppleShare 4.0.2 or later. This means that the 9150's performance as a fileserver will lag behind the AWS 95 until improved server software becomes available.

Performance

Determining performance on the Power Macs is not particularly straightforward. Emulated software receives the benefit of toolbox acceleration, introduced in Chapter 1 and discussed further in Chapter 3. Native software can be slowed down by the parts of the operating system that are still emulated. In the meantime, Apple's engineers are working on making more of the operating system native. As a result, these are the first Macs that will gain significant performance over time just by installing new software.

With native software, there is little question that the Power Macs best the 68k-based Macs in price versus performance.

If you're interested in a greater degree of detail than this chapter provides, see Chapter 7.

CHAPTER THREE

Power Macintosh Software Overview

he Power Macintosh hardware is impressive, but the success of the Power Macs rests squarely on how well software performs—both native and emulated software. Without blazingly fast native applications, a Power Mac is just another Mac and certainly not competitive with offerings from the 80x86 world. With native apps, however, even the first-generation Power Macs offer performance rivaling that of some workstations, but with the familiar and easy-to-use Macintosh user interface—no UNIX required.

Power Macintosh System Software

The Power Macs run System 7 by default—System 7.1.2 and later, to be precise. They are real Macs, not disguised UNIX workstations, nor do they use the Taligent operating system, which won't ship until 1995.

The same System 7 environment that existing Mac users are familiar with is the default operating system for these new Macs. Although the new version of System 7 contains enhancements to run faster on Power Macs, there are no visible changes to the Macintosh interface as a result of installing the new version of the Mac OS.

Installing System 7.1.2

There are two ways to install System 7.1.2: either as a fresh installation or as an update to an existing system for 68k Macs. The latter option is easier for users of existing Macs who want to move to the Power Macs with a minimum of hassle, but a fresh installation reduces the possibility of mishap during installation.

If at all possible, you should perform a fresh installation of a new version of the Mac OS and copy all the third-party software and preferences in your System Folder and its subfolders into the freshly installed System Folder.

The only instance where an update might be of greater convenience is when installing on a Mac with a Power Macintosh Upgrade Card. However, since the full installation from the Upgrade Card's system-software disks will also run in 68k mode Macs that accept Upgrade Cards, the benefit is negligible when compared to the security afforded by a freshly installed system. Furthermore, you're assured that with a fresh installation, your system software contains all the latest versions of network drivers and similar software that sometimes get missed when installing over an existing system.

Basic Power Macintosh OS Features

System 7.1.2 has the same basic features as System 7.1, with additional enhancements for the Power Macs. The version number may be somewhat misleading: System 7.1.2 doesn't contain all the features that came with System 7.1.1, which was the system-software release included in System 7 Pro. PowerTalk and AppleScript, the main additions to System 7.1 for the Pro version, are still separate parts of system software and aren't automatically installed with 7.1.2.

Anyone upgrading a System 7 Pro system to run on a Power Mac must also install an update to PowerTalk, version 1.0.3 or later, to run on the Power Macs. The 1.0 version of

PowerTalk that comes with System 7 Pro will not run on the Power Macs.

However, the standard installation of 7.1.2 automatically installs QuickTime 1.6.2 and its accompanying QuickTime PowerPlug file, Apple's CD-ROM driver 4.5, AppleTalk 58.1, and EtherTalk 2.5.5.

Toolbox Acceleration

Apple's system-software engineers have understood the concept of toolbox acceleration since the release of the 8•24 GC card in 1989. This video card came with an AMD 29000 (29k) RISC processor installed, and it was running a version of QuickDraw on the 29k. Anytime the host Macintosh used QuickDraw, the code on the GC card would be executed, much faster than the 68k code on the host Macintosh. Thus was born toolbox acceleration, the selective replacement of performance-critical parts of the Macintosh operating system with faster software versions running on faster hardware. Following the 90/10 rule, where approximately 10 percent of the code is executed 90 percent of the time, the acceleration of QuickDraw provided a disproportionately large performance improvement, since all Macintosh software uses QuickDraw in one way or another. Making QuickDraw run faster made all software run faster.

When Apple set out to develop the system software for the PowerPC-based Macs, some engineers wanted to make the entire system-software release for the new machines native. That laudable goal was soon proven to be overly optimistic. A second strategy—toolbox acceleration—was adopted. Typical applications were profiled to determine in which parts of the operating system the most time was spent. The more time spent in a routine, the more important it was to make it native.

Using the data gathered by the profiling, Apple's engineers made decisions about which parts of the operating system to make native, and which parts of the operating system wouldn't gain from being native. QuickDraw was found to be important for all Mac software, so it was made native in its entirety. Other frequently used parts of the operating system, such as parts of the Resource and Memory Managers, were also found to provide a benefit to almost all software, so they were made native as well. Throughout the development of the first version of system software for the Power Macs, a clear performance benefit was necessary before the engineers decided to make a particular part of the OS native. Since time and resources were limited at Apple's end, every bit of native code had to count.

Emulation

The emulator in the Power Macs' ROM, which is discussed in greater detail in Chapter 6, is the cornerstone in the Power Macs' system software. Without the emulator, existing 68k Macintosh software couldn't run on the new PowerPC-based machines. But equally important is the fact that much of the Power Macs' operating system is still 68k code. It isn't all native yet, so the emulator is necessary for the Mac OS to run on these machines.

Consequently, the Power Macs' 68k emulator must be extremely reliable and compatible to be able to run all of Apple's 68k system software as well as all the third-party 68k software that users already have.

The 68k emulator in the Power Macs acts like a 68LC040 running in user mode. The 68LC040 is a variant of the 68040 that lacks a floating-point unit on the chip. The user-mode distinction is an important one, because the emulator doesn't support any of the 68040 instructions that control the memory-management unit. These instructions are supervisor-mode instructions on any 68k processor, as well as on any PowerPC processor. Since only the system software should be using memory-management-unit instructions to begin with, this is not a real limitation for the emulator. Also, the 601's memory-management model is different from that of the 68k family, so emulation wouldn't make much sense anyway.

The Macintosh Centris 610 uses a 68LC040, as do the PowerBook 500 series machines, the 520, 520c, 540, and 540c; they also lack floating-point hardware support, and the few pieces of software that depend on floating-point hardware will not run on these 68k-based Macs either. However, any software that uses SANE (Standard Apple Numerics Environment) will continue to run on the Power Macs, since SANE is supported, but only for emulated software. Native software should use the PowerPC's native floating-point support directly to take full advantage of the performance available.

Native PowerPC System Software

Most of the system software for the Power Macs still consists of 68k code, which runs under emulation on the Power Macs. However, several parts of the operating system that are executed most frequently are native on the Power Macs. Some of these pieces of system software are in the ROM, some in the System file, some in the system enabler for the PowerPC Macs, and some in separate files.

Regardless of their location, these bits of native code benefit both native and emulated software. Emulated software gains particularly since, by definition, an emulator runs slower than hardware. When emulated applications call the operating system and the part of the OS that is being called is native on a Power Mac, the emulated application gets the full advantage of the native system software without needing any specific knowledge of it. For that reason, emulated software should never need to know whether a particular part of the OS is emulated. Some native software needs to know such details for performance reasons, but as a rule, most software shouldn't care whether parts of the operating system are native, since access to them from the software's perspective is identical.

Mixed Mode

At first glance, mixing emulated 68k code and native code seems to be a tricky endeavor. It is. However, this complication is completely shielded from emulated 68k software thanks to a new piece of system software called the Mixed Mode Manager. Since emulated software has to be able to run on the Power Macs unchanged, the combination of the emulator and the Mixed Mode Manager lets emulated 68k software run the same way as on a 68k-based Mac. See Figure 3.1.

The Mixed Mode Manager keeps track of the type of code being executed at the moment. It knows when native code is running, and it knows when code is running under emulation. The important task for the Mixed Mode Manager to perform is the orderly transition between the two. Since much of the Power Macs' operating system is still emulated, native apps need to be able to call emulated code. The converse is true as well: Emulated apps automatically benefit from native QuickDraw, so emulated code must be able to call native code. The most amazing part about this is that the two types of code don't even need to know about each other.

Despite the magic of mixed mode, there is a downside to being able to execute two different types of code without a hiccup. The transitions between the two modes are slow. In fact, frequent transitions between 68k and native PowerPC code can negate the performance benefits offered by native code. Such transitions are referred to as *mixed-mode switches*.

FIGURE 3.1
Mixed Mode Manager

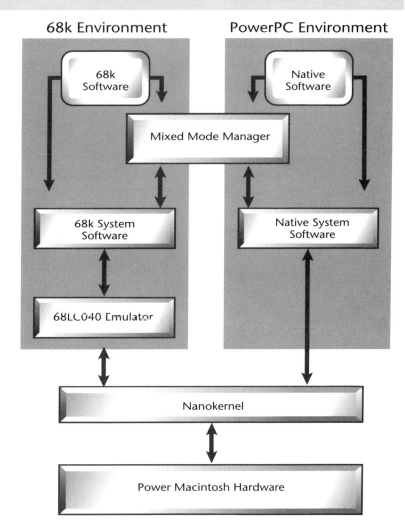

In some cases, native software can cause a slowdown because of the mixed-mode switches involved. See Figure 3.2.

The speed improvement of native PowerPC code on the Power Macs is not necessarily a given: When emulated software calls on native software to perform some work, the operating system has to do some housekeeping during each mixed-mode switch, before and after the switch from

FIGURE 3.2
Mixed-mode Switches

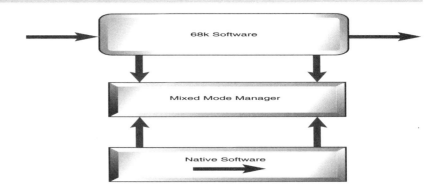

running emulated to running native and back. Sometimes, these housekeeping chores can take longer than the time gained by running native, particularly if the operation to be performed by the native software is very short. So it might seem productive to switch from 68k code to PowerPC code wherever possible, because native software executes as much as 10 times faster. But staying in emulated 68k code can sometimes be faster than switching if the time saved is more than the time lost due to two mixed-mode switches.

Understanding mixed-mode switches is important to further understand the decisions of developers at Apple and elsewhere regarding which parts of code they chose to make native. In many cases, native code isn't faster and would simply cause mixed-mode switches; in such cases, executing emulated code winds up being faster. If the entire Mac OS were native from top to bottom, and all third-party products, including INITs, drivers, and other low-level pieces of software, were as well, then there wouldn't be a problem. However, making the entire operating system for the Power Macs native would have delayed the release of the Power Macs and would also have caused significant compatibility hassles, especially with third-party extensions.

Since many parts of the Mac OS aren't native, it doesn't make sense to make certain kinds of software—SCSI drivers,

SCSI Manager 4.3

Chapter 2 provided an overview describing the new SCSI hardware and the DMA supporting it. However, this new hardware wouldn't be able to run at its highest speed without the appropriate software. The Centris 660AV and the Quadra 840AV were the first Macs with DMA SCSI that was usable from the Mac OS. (The Macintosh IIfx has the dubious distinction of being the first Macintosh with SCSI DMA, but only users of A/UX, Apple's version of UNIX for Macs, were able to benefit from it.) To be able to take advantage of this new SCSI hardware, the new SCSI Manager 4.3 is required. It, too, was first introduced with the 660AV and 840AV.

The new SCSI Manager provides several benefits over the old SCSI Manager: It has support for DMA, it supports asynchronous I/O as well as SCSI features such as disconnect, and it has support for multiple SCSI buses.

However, support for DMA does not mean that DMA is required for the new SCSI Manager to be active. A Quadra with an installed and active Power Mac Upgrade Card also has SCSI Manager 4.3 installed, but no DMA. The DMA hardware in the Power Macs allows the new SCSI Manager to take fullest advantage of the available performance in the SCSI hardware. Because of the way SCSI Manager 4.3 is designed, it can set up a SCSI transaction and let the DMA hardware do all the work. The SCSI Manager gets out of the way once the transaction is started.

Asynchronous I/O, a term that's applicable to the Macintosh operating system, shouldn't be confused with asynchronous and synchronous SCSI transactions. Async I/O in the Mac OS sense means that software—either application software or any other kind of software that's likely to cause a SCSI transaction—can call the OS to do its bidding, for example to write data to a hard disk, and while the SCSI Manager is off doing that, the software can go on and perform other work while the SCSI I/O is taking place. Before SCSI Manager 4.3, the OS wasn't able to support such simultaneous SCSI activity.

for example—native, since such software would cause many mixed-mode switches and not enhance performance at all. On the contrary, a native SCSI driver would slow down performance. Such a driver isn't beneficial until a native SCSI Manager as well as a native version of the Mac's hierarchical file system is available.

One way that some low-level software can avoid unnecessary mixed-mode switching is by being *fat,* a term used to describe software that contains 68k and PowerPC code. In the case of low-level software that needs to hook into parts of the operating system, installing so-called fat patches reduces the number of mixed-mode switches. The reason for this is that the Mixed Mode Manager always tries to stay in the mode it's currently in. So if emulated software is executing and it calls a piece of the operating system that has been augmented by a fat patch from third-party software, the Mixed Mode Manager executes the 68k version of the patch software to avoid a mixed-mode switch. As with the preceding driver example, the only time when it makes sense to provide native software is if the native version, including mixed-mode switches, is always faster. In such cases, any software that installs patches can install only a native patch on a Power Mac, since all software will profit from the greater speed.

If you'd like to learn more about the Mixed Mode Manager, it is discussed in greater depth in Chapter 8.

Native QuickDraw

QuickDraw on the Power Macs is entirely native—no emulation anywhere. The reason for this is straightforward: In all cases, even when calling it from emulated apps and causing a minimum of two mixed-mode switches, Native QuickDraw (NQD) runs faster than emulated QuickDraw.

As part of the investigations into the parts of the Macintosh operating system where the most time was spent, QuickDraw came out as one of the clear leaders. Since the Macintosh is a graphics-oriented system, this is no great surprise.

Native QuickDraw is an evolution of the version of QuickDraw present in the Quadras, version 1.3.0. This version of the Mac's imaging system software is still largely written in 68k assembly language for performance reasons. NQD, whose version number is 1.3.5, is based upon version 1.3.0 but rewritten completely in C and compiled for PowerPC.

Intuitively, one might think that a C-based version of QuickDraw would be significantly slower than a hand-tuned one written in assembly language. This may be so on a CISC processor, but it isn't necessarily the case with a RISC processor. Many of the performance benefits of RISC processors require sometimes fiendishly clever machine-language constructions; a compiler can take human-legible and -maintainable C code and translate it into highly optimized RISC machine language that runs extremely fast. In the CISC days, a good assembly-language programmer could almost always write better code than a compiler. With today's complex RISC chips, however, the compilers often generate faster code than handwritten assembly. For this reason, one of the translations of the RISC acronym is "relegate the interesting stuff to the compilers". If you'd like to learn more about how RISC chips and the PowerPC family in general work, see Chapters 4 and 5.

Native QuickDraw runs many times faster than QuickDraw 1.3.0. It not only benefits from running on a faster microprocessor, but NQD also takes advantage of specific features of the PowerPC family to boost performance even further. Some of the most processor-intensive parts of NQD take advantage of the PowerPC's 64-bit-wide data bus to write graphics data out to memory as quickly as possible.

Both emulated and native apps benefit from Native QuickDraw's speed. Those emulated apps that use QuickDraw frequently show especially large speedups when running on a Power Mac. Much like Apple's 8•24 GC card, which contained its own RISC processor and a RISC version of QuickDraw, the Power Mac's NQD acts like a QuickDraw accelerator. Unlike many hardware accelerators that accelerate only the most time-critical parts of QuickDraw, such as those that move large blocks of a screen around, all of NQD is accelerated.

Since all of QuickDraw is native on the Power Macs and since it's used so frequently, NQD's performance is particularly susceptible to slowdown from excessive mixed-mode switches. Extensions that install patches into QuickDraw, either to enhance features or to accelerate it in a 68k environment, can cause NQD to be throttled down to significantly lower performance. The section on third-party software later in this chapter discusses this issue in greater depth.

Native QuickTime

The Power Macs come with QuickTime 1.6.2, which by itself consists exclusively of 68k code. When you install QuickTime on a Power Macintosh, however, you also install the QuickTime PowerPlug, which contains native versions of the most processor-intensive parts of QuickTime. Much of QuickTime is dependent on the speed with which it can get data to and from where it needs to be. The frame rate of QuickTime playback is determined largely by the speed of the hard disk or CD-ROM drive that the QuickTime data is on, but also by the speed of the video hardware and software. Since the Power Macs' I/O features are emulated, and the hardware's SCSI and Ethernet DMA do the work of getting the data from mass storage to memory, the emulated parts of QuickTime are less critical than one might think. The same logic that was used for toolbox acceleration is used in QuickTime: Only the parts where the most time is spent, or those parts that would benefit the most, are made native.

CPU-intensive parts of QuickTime such as the Cinepak compressor/decompressor (codec), as well as other codecs, are native and come in the QuickTime PowerPlug.

QuickTime version 2.0 follows the same philosophy as 1.6.2 with its native support. QuickTime 2.0 has inherent performance improvements as a result of changes made to some of its internal operations—for example, the new data-pipe I/O architecture within QuickTime 2.0 provides a big performance boost by itself, even though it is not native. This new version of QuickTime also comes with its own QuickTime PowerPlug that, like 1.6.2's, contains the most performance-critical parts, the compressor/decompressors (codecs).

Memory: Modern and Virtual Both

Another critical part of the revised operating system for the Power Macs is the Modern Memory Manager, which is in charge of allocating and deallocating parts of memory. The Modern Memory Manager is a complete rewrite of a venerable part of system software that's been with the Macintosh since 1984. Despite being a complete rewrite, the Modern Memory Manager behaves just like the old ones; typical software need not change to use it and take advantage of it.

The Modern Memory Manager makes itself known to users only by snappier Macintosh performance. That and the additional items in the Memory control panel that allow the Modern Memory Manager to be turned on and off are the only outward manifestations of this new system software. Internally, the Modern Memory Manager does the same job as the old Memory Manager, only more efficiently and quickly. Since all software on the Mac uses the Memory Manager, a faster one benefits the entire system.

The Modern Memory Manager is fat: It exists both as 68k and as PowerPC code. Since mixed-mode switches can cause such a slowdown, it's beneficial to avoid them where possible. A fat memory manager means that native software calls a native memory manager and causes no mixed-mode switch, and emulated software calls an emulated memory manager, likewise without a mixed-mode switch.

In addition to the modern memory manager's greater basic efficiency, it coexists with virtual memory (VM) much better. When either of the memory managers receives a request from software that wants to have access to another block of memory, the memory manager must first find a block of RAM to allocate to the software. The previous memory manager would go looking around in RAM to see where it could find a block of unused RAM; this hunt for RAM happened without regard for whether a part of RAM was swapped to disk or whether it was really in RAM. This behavior caused frequent *page swaps,* and performance deteriorated as a result. The Modern Memory Manager is aware of which parts of RAM are real RAM and which have been temporarily stored to disk. By keeping track of this information, the Modern Memory Manager doesn't cause any unnecessary swapping and thus keeps performance high.

But why use virtual memory in the first place? On the 68k Macs, this is a reasonable question. VM is much slower than real RAM, real RAM is relatively cheap, and those applications that really need lots of RAM—such as Adobe Photoshop—don't operate well with VM enabled. But on the Power Macs, VM has a memory-saving benefit that makes the use of existing RAM more efficient, even if the Power Macintosh has plenty of RAM installed.

A large drawback of virtual memory on the Macintosh, even on the Power Macs, is that it must create a VM swap file on a local hard drive that is equal in size to the total amount of memory available with VM enabled. For example, let's say your Mac has 16MB of real RAM, and VM is configured for the smallest size—1MB in addition to available RAM. When you turn on your Mac, the VM system software creates a 17MB file on the volume that you specify in the Memory control panel. There is no way around this. The more real RAM you have, the larger the swap file becomes if you enable virtual memory.

On the Power Macs, there is a good reason to enable VM, especially if you use many native apps. The executable code for native apps is stored in one contiguous piece in the data fork of an application file. Previously, the data fork of applications has been unused by the operating system. When virtual memory is enabled on the Power Macs, it treats the area on disk where a native app's executable code resides as a type of VM swap file. Only the application code that's really needed is loaded into RAM, and code that's needed later on is transparently loaded into RAM by the VM system software. Another side effect of all this is that with VM enabled, native code is protected by the PowerPC chip's built-in memory-management hardware. Native code is marked as read-only in RAM, so anything that tries to write to a part of RAM that contains code will be thwarted. This is a small, early step toward a Macintosh operating system with memory protection. Read-only code has another benefit as well: Since this code can never be modified, there's never any need to write it back to disk if the memory it was occupying is needed. So, instead of writing the code to disk before reading in other code, the VM system software need only read in the new code, saving a time-consuming write to disk.

I/O

All of the operating-system code that handles input and output, from and to the various ports inside and outside the Power Macs, is emulated in this version of the operating system. Although this may seem silly at first, it turns out that making the I/O code native wouldn't have provided that much of a benefit. The main reason that I/O still runs in emulation on the Power Macs is compatibility. The Mac system software's methods of handling I/O rely on behaviors of the 68k microprocessor family. The emulator and the nanokernel, the lowest-level part of system software on the Power Macs, collude to make it appear as if the Power Macs had the same I/O behavior as previous 68k Macs.

The SCSI Manager 4.3, discussed earlier in this chapter, is emulated. SCSI performance gains much more from the DMA hardware than from the system software. The drivers for the serial and Ethernet ports are the same way: Once a transaction is under way, the hardware handles most of the work.

Some I/O-related system software would, however, benefit from going native, since it's fairly computationally intense.

- The AppleTalk protocol stack, as well as MacTCP, which is the TCP/IP protocol stack for Macs, handles the processing of network traffic to and from the Ethernet and LocalTalk ports of a Mac. They spend a lot of their time decoding and encoding packets and would almost certainly benefit from going native.
- Apple Remote Access, which performs error correction, compression, and protocol processing, would also benefit from being native, since today's high-speed modems can move data back and forth very quickly, and an emulated ARA has a great deal of work to do.
- All of the GeoPort software except the software modem, also known as the data pump, runs in emulation, including the error-correction and compression code. Since this is time-critical software that does a lot of processing, it would benefit from being native. However, because this software relies upon the behavior of the I/O in 68k Macs, it would be difficult to make the switch without better support within system software for native drivers and other I/O-related software.

All of these examples run well under emulation today, albeit not as fast as they could. Some of this system software is already slated to be native by the end of 1994; others have no announced plans to go native. The AppleTalk and

TCP/IP protocol stacks, for example, will be native by the end of 1994 as part of the new OpenTransport network system software. This will allow high-performance networking for both Power Mac servers and clients.

But to bring all I/O software native would have required making the Device Manager and many other parts of the OS native. Since the Device Manager is also tied to the 68k hardware design, and since a native Device Manager with support for the PowerPC would have to behave differently and would require native drivers as well, this scenario was dropped. Full native I/O support is expected when the microkernel version of the Macintosh operating system is introduced.

INITs and Patches

Users can customize their systems to their heart's content by dropping extensions and control panels into their System Folder. A great deal of nonfrivolous software also requires the installation of an extension or a control panel. In the past, a slowdown has always been associated with using many extensions and control panels, but this slowdown was rarely severe enough to worry about. With the advent of the Power Mac and its emulator, extensions and control panels still work. However, some of these extensions and control panels install code into your system that can drastically decelerate your Power Mac.

Mixed-mode switches are again the problem here. Many of the extensions and control panels install *patches,* which replace or reroute existing system code. Such patches can cause performance degradation by themselves, but coupled with a mixed-mode switch, the performance loss can be great.

On the Power Macs, you can install a 68k patch, a PowerPC patch, or a fat patch that contains both 68k and

PowerPC code. The best type of patch for a given situation depends on the part of system software that's being patched and how much computation it performs. However, system software consisting of PowerPC code should never be patched with 68k code, since in all cases, this will cause a performance hit and slow down the system. The severity of performance loss in this situation depends on how often the patched code is executed. If it's called frequently, the performance loss will be great. Adobe Type Manager (ATM), for example, was not available in a native version when the Power Macs were introduced. Apple's profiling had, however, identified text-drawing as the single most time-critical part of the operating system; for this reason, the text-drawing code in system software was PowerPC code. Since ATM installed a 68k patch into a frequently called PowerPC routine and caused many mixed-mode switches, overall system performance decreased measurably. In one test performed with a word processor, the presence of ATM caused as much as a 30 percent performance degradation when scrolling. Unfortunately, since so many people are dependent on ATM to provide high-resolution Type 1 fonts, most Power Mac users were stuck with a decelerating ATM until Adobe released a native version with native patches.

Fat patches are discussed further in Chapter 8.

GeoPort for Power Macintosh

Although GeoPort is a hardware feature of the Power Macs and uses an external adapter to connect to phone lines, it requires software to work as a modem. The GeoPort software for Power Macintosh is native code that performs all the work of a traditional hardware-based modem exclusively on the PowerPC processor.

A GeoPort modem is made up of several software parts. At the lowest level is the driver that allows the Power Mac to communicate with the external GeoPort Telecom Adapter at 2 megabits per second. This high data rate is necessary to transmit and receive the audio data between the phone line and the software modem in the Power Mac. The GeoPort Telecom Adapter is a straightforward piece of hardware that converts the analog audio data from the phone line to digital data for transfer to the Power Mac.

The next part is the native software that performs all the signal processing and is the actual modem. The PowerPC processor family supports a particular instruction that is the core operation performed by dedicated digital signal processor (DSP) chips. The PowerPC 601 in the Power Macs can therefore do much of the same work that a dedicated DSP chip can, but the Power Macs need not incur the additional cost of adding dedicated DSP hardware to the system. At this writing, the native software modem supports data connections at up to 14400bps (bits per second) using the V.32bis modem standard, and fax connections of up to 9600bps via the V.29 standard.

Layered on top of the driver and the software modem is Apple's Express Modem software, which is the part that application software interacts with. The Express Modem software contains an AT command interpreter. AT commands are the standard method of configuring a modem, and most communications software relies on being able to send a modem AT commands, so a good software modem must support them as well. The Express Modem software also contains code that provides standard error-correction protocols such as V.42 as well as standard data-compression protocols such as V.42bis.

The GeoPort software simulation of a hardware modem is well-rounded and has no fundamental omissions; communications software has no idea that it's not dealing with a piece of hardware. Compared with a hardware modem

> ### Compatibility
>
> Compatibility means different things to different people, but it's clear that compatibility is a good thing in everyone's book. In the past, moving from an older Mac to a newer one, or upgrading to the latest version of the OS, generally brought problems, and users had to upgrade some of their third-party software to work with the new stuff. This time around, things are different. Most everyone is so intent on seeing how compatible the new Power Macs are, and expectations are higher than they would be for any other Mac.
>
> Hardware and software compatibility issues cropped up with the introduction of the Quadra 660AV and Quadra 840AV, but since these machines were considered high-end, fewer people than usual encountered these problems. Many of the same issues that proved to be compatibility problems with the AV Quadras can also be problematic with the Power Macs. On the hardware side, Chapter 2 illustrates how the increased SCSI performance of the 660AV, the 840AV, and the Power Macs also results in a more finicky SCSI bus. Similar issues face users of existing Mac software.
>
> The emulator itself is remarkably solid, but if you have software that's a bit older, especially extensions and control panels, you should make sure you have the latest versions before making the switch from 68k to PowerPC. This goes not only for third-party software, but also for Apple software that isn't part of the operating system—Apple Remote Access, for example. ARA 1.0 will not work on the Power Macs, but the ARA 2.0 client software, which was released well before the introduction of the Power Macs, works fine.

with a similar feature set, the GeoPort modem for Power Macintosh is inexpensive. The software is free, and the GeoPort Telecom Adapter to connect the Power Mac to phone lines is cheaper than a V.32bis modem.

Networking Software

Since the networking hardware in the Power Macs is identical to that in the Quadra 660AV and Quadra 840AV, and the AppleTalk and EtherTalk software is emulated, existing network software can be used on the Power Macs. In addition, as Apple comes out with newer versions of AppleTalk as well as newer LocalTalk and Ethernet drivers, these new versions can also be used on the Power Macs. Some of Apple's own installers don't mention the Power Macs by name in the Installer options yet, but network software intended for the 660AV and 840AV is suitable for the Power Macs as well. Other networking software, such as the AppleShare client and most third-party network software, runs without mishap on the Power Macs.

Some networking apps, such as the protocol analyzers from Neon Software and the AG Group, benefit from going native despite the Ethernet drivers being emulated. This software is used to analyze large amounts of network traffic, usually to track down a problem with the network. The faster the software can crunch through the captured packets and figure out what's going on, the faster the user can get on with fixing the problem. Although this isn't a typical Power Mac application, it does illustrate how the PowerPC's raw performance can boost the productivity of users whose software is limited by low-level parts of the OS that are still emulated.

Apple Business Systems Software

Apple Business Systems' software runs on the Power Macs, but none of the major products from ABS will be native before the end of 1994. The AppleShare server software was upgraded to version 4.0.2 to add support for the Power Macs, desktop models as well as the Apple Workgroup Servers 6150, 8150, and 9150. This new version of AppleShare contains changes made for compatibility reasons as well as some performance enhancements. The server software is still emulated, so these changes bring the performance of Power Mac AppleShare servers up to and, in some cases, beyond the performance of AppleShare servers running under the Mac OS on 68k hardware. In the future, as more of the server software available for Macs goes native, and as the OpenTransport native protocol stacks also become available, the PowerPC-based Mac servers will get major performance boosts just from software upgrades.

For the fastest possible server performance, NetWare running on a Power Mac and AppleShare Pro, which runs under A/UX, will remain the kings of the hill. Apple's AppleShare servers are designed primarily to serve workgroups, and these high-performance server packages are designed with large workgroups and multiple departments in mind. They also cost accordingly.

Floating Point—Who Needs It Anyway?

Much ado has been made about the Power Macs' floating-point capabilities, but a big question remains: Who, other than 3D renderers, cares? Without software that takes advantage of the screamingly fast hardware, it doesn't provide the user with any tangible benefit.

Until the introduction of the Power Macs, the only software that used floating-point hardware was relatively specialized: 3D software, scientific and engineering software, and Fortran compilers all took advantage of the 68k floating-point hardware if it was available. Many such applications even required it, since without hardware support, the software would be unusably slow. However, the performance boost provided by taking advantage of the 68k floating-point hardware wasn't enough for mainstream developers to make the effort to change their code to use it. In addition, floating-point hardware wasn't available in all 68k Macs, so the work required to use the floating-point hardware would benefit only a subset of the buying populace.

In contrast, floating-point hardware support is part of the PowerPC architecture specification. Any PowerPC processor must be able to handle floating-point instructions. This

> ### NetWare on Power Macintosh
>
> Apple and Novell announced in April 1994 that they are collaborating on the development of a NetWare 4.1 port for the Power Macintosh hardware. As with NetWare implementations on x86 processors, NetWare is the OS for such a machine; it does not run with another operating system the way AppleShare runs on top of the Macintosh operating system.
>
> NetWare on the Power Macintosh will look and feel just like any NetWare 4 server. Its management user interface will be identical to other implementations of NetWare 4—no Macintosh front end. You also won't be able to run any Macintosh software on a Power Mac NetWare server, since NetWare takes over the machine completely. The Power Mac version of NetWare comes with the NetWare for Macintosh server software preinstalled. This NetWare Loadable Module is required to allow Macs using the AppleShare client software to connect to a NetWare server.
>
> In the future, Novell plans to offer a NetWare client that uses Novell's de facto standard IPX/SPX protocols. Novell has already shipped MacIPX, a 68k Macintosh implementation of the IPX/SPX protocol stack, but it has not yet provided a NetWare client that uses MacIPX. There are also plans to deliver a native IPX/SPX protocol stack for the OpenTransport architecture that Apple will introduce in late 1994. Native IPX will allow the highest performance for Power Macintosh clients connecting to NetWare servers.
>
> The Power Mac version of NetWare will not replace AppleShare; where AppleShare is focused on workgroups with tens of people, NetWare 4 is designed with hundreds of users in mind. NetWare on Power Macintosh will consequently cost much more than AppleShare.

means that every Power Mac will always have hardware floating-point support. However, Power Macs do not include emulated support for 68k floating-point hardware: 68k apps that require 68k floating-point hardware will not run on the Power Macs. The only high-performance floating-point support on the Power Macs is for native apps.

This doesn't mean that emulated apps can't perform floating-point operations: SANE is supported on the Power Macs. SANE has been in the ROM of every Macintosh since the original 128k Mac. Its purpose was to offer highly accurate and consistent floating-point results on all Macs. At the time, the 68020 and its floating-point sidekick the 68881, didn't exist yet. SANE made these floating-point features available by performing the calculations with integer operations, a slower but equally effective way of going about this. In fact, SANE later turned out to be more accurate in some calculations than Motorola's own floating-point chips.

SANE on the Power Macs is also implemented exclusively using integer code. One reason for this is that SANE's main floating-point number format, the 80-bit large *extended-precision* format, is different from the 64-bit double-precision format used by the PowerPC family. If SANE were to use the floating-point hardware in the PowerPC, a great deal of time would be spent converting between the 64- and 80-bit formats. This frequent conversion would have a big impact on performance. More important, though, the calculations performed by a PowerPC hardware–assisted SANE would be less accurate, since it would be using only the 64-bit numbers rather than 80-bit numbers. For this reason, SANE is native on the Power Macs, but it performs all calculations with 80-bit precision using integer code. As a result, SANE is still considerably faster than SANE on any 68k-based Mac, but it's nowhere near as fast as the somewhat less accurate native PowerPC floating-point hardware.

SANE is an Apple-defined standard that wasn't adopted by any other computer vendor, since, in many ways, it was well ahead of its time. Comparable floating-point standards exist today, and the native PowerPC Numerics environment on the Mac supports them. IEEE 754 is the name of the standards definition put forth by the IEEE and NCEG.

UNIX and the Power Macintosh

Even though UNIX isn't important for most personal-computer users, some users need to run UNIX on occasion.

Apple has traditionally made UNIX available for the Macintosh in the form of A/UX, which evolved through three major revisions to be a working hybrid of the Macintosh operating system and UNIX. However, A/UX runs only on 68k-based Macs. At the introduction of the Power Macs, no mention was made of UNIX support for the new PowerPC-based machines. Since then, the UNIX picture has cleared up somewhat. The Power Macs will ultimately be able to run two versions of UNIX.

Tenon Intersystems' MachTen is a version of UNIX based on the Mach microkernel. MachTen is a novel approach because it runs as an application under the Macintosh operating system rather than taking over the entire Mac for itself. Within its own environment, MachTen provides preemptive multitasking, virtual memory, and other standard UNIX features. But it also behaves like a Macintosh application. This allows Mac apps and UNIX apps to be running on the same machine at the same time, with neither aware of the other.

MachTen runs emulated on the Power Macs, except that it doesn't provide virtual-memory capability, since the emulator doesn't emulate a 68k MMU. Otherwise, MachTen behaves like it normally does. Tenon intends to ship a cross-development kit in the third quarter of 1994 that allows the creation of PowerPC native MachTen binaries. In the fourth quarter of this year, Tenon plans to ship a native version of MachTen that still behaves like a Macintosh application, but performs much better and has support for native MachTen apps.

The other version of UNIX available for Power Macintosh will be based on IBM's AIX 4.1. This version of UNIX is compliant with the PowerOpen specification, which defines a standard operating environment for UNIX software running on PowerPC-based systems. The idea behind PowerOpen is to allow PowerOpen-compliant applications to run on any PowerOpen-compliant operating system. If Apple and IBM ship different versions of UNIX, but both are PowerOpen-compliant, software that runs on one should run on the other as well.

PowerOpen also includes software known as Macintosh Application Services. MAS allows the user to run 68k Macintosh applications on a PowerOpen-compliant system, just like A/UX did. MAS includes a 68k emulator, albeit a different one than in the Power Mac's ROM, as well as an implementation of the Macintosh operating system that translates many of the Mac OS calls into UNIX calls, completely transparently to the Macintosh software.

At this writing, it was unclear when a PowerOpen OS would be available for the Power Macs, or when IBM's AIX 4.1 would ship.

Software on the Power Macintosh

The combination of PowerPC-native and emulated 68k software on the Power Macs makes the software environment on these new machines significantly more complex than on previous Macs. In many instances—I/O software being a prime example—the intuitive conclusion that native software is automatically better is a false one. Emulated system software still has good reasons for existing.

The Power Macs' selectively native system software accelerates performance for the most commonly used routines in the OS. Many parts of the operating system and toolbox still aren't native and have no need to remain emulated the way I/O software does. These parts are likely to go native over time and be provided by Apple as incremental performance enhancements. The Power Macs will get faster over time simply by adding new software.

Users are also finding certain bottlenecks in system software that aren't necessarily easy to predict, no matter how much profiling is done. Any task that uses Copy and Paste frequently from within native software will find it slow going; the frequent mixed-mode switches going between the native application and the emulated clipboard code can act as decelerators. This type of issue will be dealt with over time as more of the Power Macintosh system software goes native.

At this point, it looks like Apple's decision to favor compatibility over performance has served it well. The uproar over compatibility issues would have been far worse than the discovery of limited performance problems such as the emulated clipboard. Thanks to the high compatibility afforded by the first generation of PowerPC-based Macs, it appears that the migration from 68k to PowerPC is well under way. The next step in the process is to concentrate on allowing native software to reach its fullest performance potential, but this will require significant amounts of native system software, which will take time. Until then, the Power Macs work, and those computationally intensive apps that run native get the majority of the performance boost today, with the promise of even more to come later.

Macintosh Application Environment (MAE)

UNIX users with SPARC or HP PA-RISC hardware can also run 68k Mac apps with the help of the Macintosh Application Environment for these two platforms. MAE is a product developed by Apple; the core technology is the same as for MAS for PowerOpen.

MAE is available for SPARC-based workstations running Solaris 2.3 or later, and for Hewlett-Packard PA-RISC-based workstations running HP/UX 9.0 or later. MAE is a separate UNIX process that mimics a 68k Mac. It translates many Macintosh OS calls to their UNIX counterparts, and it also translates the Macintosh QuickDraw graphics calls into X Windows commands.

MAE runs in a window on the host workstations, just like any other UNIX application. It has a Finder and supports printing. AppleTalk-based networking is not part of the first release, though; traditional methods must be used to move files back and forth between Macs and a workstation running MAE over the network.

CHAPTER FOUR

An Introduction to Microprocessors

The term *microprocessor* refers to a type of integrated circuit, or chip, that is designed and used to perform processing of some kind, primarily calculation. RAM chips, for example, aren't microprocessors because they don't transform any input values into different output values; they are simply storage devices. As the need for faster processors grows, microprocessors are becoming increasingly complex, with a vast array of different features, many of them increasingly subtle or esoteric. Understanding each processor's intricate design details these days is beyond the scope of even the most interested individual. However, current microprocessors share traits that allow an adequate understanding of the processors' function and that also allow a reasonably accurate comparison of their different abilities.

The design of a microprocessor can be deconstructed into different levels: higher ones, such as a processor's architecture, and lower levels, such as implementation details that are specific to a single microprocessor.

Fundamental Microprocessor Concepts

To understand the higher-level issues and features of microprocessors, knowledge of some basic microprocessor concepts is required.

Cycle

A *cycle* is the smallest measurement of time for a microprocessor or in a computer system. All computer systems use signals generated by clocks to synchronize the different parts of the system and keep them running together. One full period of a clock signal is called a cycle.

Cycles are measured in hertz. The PowerPC 601 processor in a Power Mac 6100/60 runs at 60 megahertz (60MHz), which means that it performs 60 million cycles of work per second.

Address

To access information in memory requires knowledge of where in memory that data resides. An *address* is a numeric value, much like an address in the real world, that describes a location in memory. Each byte in memory has its own address.

A *pointer* is an address that points to specific information in memory. Pointers commonly have specific types, depending on the data that's being pointed at.

Register

A *register* is the fastest and smallest type of memory in a computer system. It resides directly on the microprocessor and is used to store data or addresses. Most operations performed on a RISC processor are performed on data and addresses in registers. In contrast, CISC processors perform many of their operations directly on values stored in memory outside of the microprocessor. Such operations take significantly more time, since accessing memory is much slower than accessing a register.

A group of registers on a microprocessor, such as the 32 general-purpose registers (GPRs) on PowerPC chips, is referred to as a *register file*. The PowerPC 601 processor has two main register files: one is made up of the GPRs; the other consists of the 32 64-bit floating-point registers.

Instruction

An *instruction* constitutes the smallest amount of work that a microprocessor can perform.

Each instruction has a unique numeric value and is stored in memory where it can be fetched by the microprocessor. The microprocessor decodes an instruction and determines the operation to be performed. In a RISC microprocessor, all instructions are the same size; in the case of the PowerPC family, all instructions are 32 bits long.

A microprocessor instruction is much like a sentence in a human language.

When an instruction is about to be executed, it is first fetched from memory, decoded, then dispatched (or issued) to the appropriate execution unit. An instruction is completed (or retired) when its result has been calculated and written back to a register.

Branch

A *branch* is a type of instruction that changes the flow of a program.

When a program is executing, individual instructions are retrieved from memory that is pointed to by a special register known as the *program counter*, which contains the address of the current instruction being executed. As each instruction is processed, the address in the program counter is incremented to point at the next instruction to be fetched. A program consists of sequentially executed instructions.

There are times when program flow must change, often because a particular condition is met. A branch instruction changes the value of the program counter to point to the next instruction it should fetch, which is not the next instruction after the branch instruction itself.

When a branch is referred to as *taken,* it means that program flow was changed as a result of the branch instruction. A branch *not taken* had no effect on the program flow.

Bus

A *bus* is a shared connection among multiple units that wish to transfer data back and forth. The most efficient way of transferring data is a direct connection between two points. But if every separate unit in a microprocessor had a dedicated connection to every other one, the overall design would be so complex that it would not be feasible to build—too complicated and consequently too expensive.

Since only a single transaction can go over a bus at a time, any device wishing access to the bus must first check whether it's free before starting a transaction; this process is known as *bus arbitration.*

Bus contention occurs when a transaction is already going on and another device connected to the bus wishes to perform a transaction as well. Since the bus is already in use, the second device must wait its turn, causing a lag.

Bus traffic describes the transactions going across the bus—both the amount of data and the time taken. The larger the amount of bus traffic, the greater is the likelihood of bus contention.

Buses exist both within microprocessors and within computer systems. A microprocessor's connection to the rest of the computer commonly consists of two buses: the data bus and the address bus. The *address bus* is used to communicate the address of the desired data to the rest of the system. The *data bus* is the pathway along which the data travels to and from the address specified by the address bus.

When data is moved across a bus, it is usually done one bus width worth of data at a time. For example, a 64-bit-wide bus can move 64 bits of data per transaction. The amount of data that is transferred during a single bus cycle is known as a *beat.*

A *burst transaction* on a bus allows a microprocessor to move a larger amount of data than usual, commonly a single cache block's worth, to or from the processor. Typical non-burst bus transactions are only as large as the bus itself—64 bits on a PowerPC 601. When moving data to or from the L1 cache, speed is of the utmost importance, so an entire cache block is moved during a burst transaction.

The speed of a burst is described in a notation that is dependent on the type of burst and the bus size; the time unit is bus cycles. For example, a 3-1-1-1 burst describes a transaction that moves four beats' worth of data; each of the numbers describes how many cycles it took to get each beat of data. The first beat takes 3 cycles because it takes time to address and access the desired location in memory. This overhead happens only once; the subsequent beats of data in the burst are moved across in a single cycle in this example.

Transistor

Transistors are the building blocks of microprocessors. They are used to construct the different functional units within a microprocessor. Transistor count is commonly used as a measurement of the complexity of a microprocessor—the more transistors used in a chip's design, the more complex that chip is. Added complexity doesn't necessarily translate into added performance, though.

Not all transistor counts are created equal, either. When comparing two microprocessors, it's worthwhile to subtract the number of transistors used in the processors' on-board cache(s), leaving only those transistors that make up the core of the processor. This is a much better indicator of a processor's complexity.

Die

Chips come into being on large circular pieces of silicon known as *wafers*. When fabrication is complete, each individual rectangular chip, known as a *die,* is cut from the wafer. The number of dice per wafer affects the price of the individual chip, since the cost of manufacturing a wafer is roughly constant. The smaller the individual die, the more dice can fit onto a wafer, and the cheaper the dice become.

Dependencies

There are two kinds of data dependencies: true dependencies and antidependencies. A *true dependency* exists when an instruction generates a new value and the subsequent instructions use that value. This is also known as a *read-after-write* dependency.

An *antidependency* exists if an instruction uses a value as an operand and the subsequent instruction creates a new value in the previous one's location. Antidependencies are also known as *write-after-read* dependencies. There is also a *write-after-write,* or *output dependency,* which is also an antidependency. This occurs when two instructions write their results to the same location, either a register or memory.

Pipelines

High-performance microprocessors achieve much of their performance through a technique called *pipelining,* in which the operations in a microprocessor's functional units are further subdivided into smaller steps, and different instructions can occur in each of the pipeline's stages. Each instruction goes through each step of the pipeline in sequence. The benefit of pipelines is that when an instruction moves from one stage in the pipeline to the next, the following instruction moves into the freshly vacated pipeline stage. See Figure 4.1.

Pipelines traditionally have four stages: fetch, decode, execute, and writeback. The first stage retrieves an instruction

FIGURE 4.1
The ideal pipeline

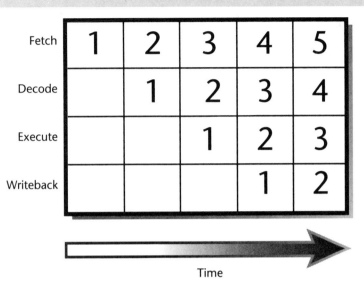

from memory or cache. The second stage decodes an instruction and fetches its operands. The third pipeline stage executes the instruction in its appropriate execution unit. The final stage writes the result of the execution stage back into the register file.

Even though each individual instruction takes multiple cycles in total, once a pipeline is full, an execution unit is able to complete an instruction every cycle. In contrast, a processor that doesn't support pipelining can issue an instruction only if the previous instruction has been completed. See Figure 4.2.

Superpipelining is a variation of pipelining where a processor's internal steps are subdivided into even more granular steps than the standard four to six pipeline stages.

Pipelines can stall. See Figure 4.3. When something hinders the pipeline from continuing at its constant pace, a stall occurs. Pipeline hazards—the factors that cause pipeline stalls—are many and varied. The most common

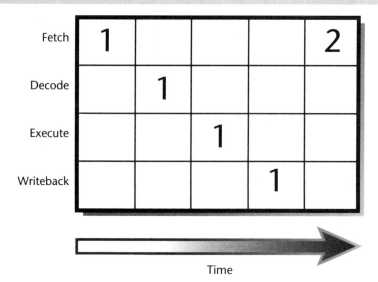

FIGURE 4.2
No pipeline

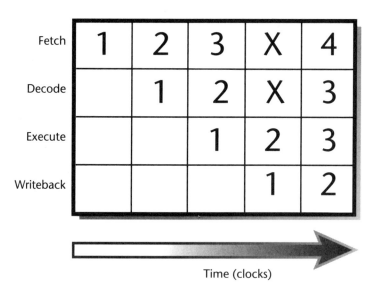

FIGURE 4.3
Pipeline stall

pipeline hazard is when two instructions are trying to access the same memory or register or if a particular stage in the pipeline takes more than a cycle to complete. When this happens, instructions following the stalled instruction must wait until it's finished, after which execution of the pipelined instructions can resume.

Since pipeline stalls break the rhythm of completing one instruction per cycle, it's worth a little extra effort to avoid stalls. RISC compilers work hard to generate instruction sequences that aren't likely to stall, but sometimes this is unavoidable.

Superscalar

A *superscalar* processor is one that can issue multiple instructions per cycle without the programmer having to think about the sequence in which the instructions are dispatched. PowerPC 6xx processors are superscalar, since all of them can issue at least an integer instruction and a floating-point instruction during the same cycle, and both instructions are processed independently of each other. Processors that require the compiler to specify multiple instructions to be issued and executed together are not superscalar; they are known as very long instruction word or VLIW.

Latency

Latency is a fancy word for wait. When trying to access memory or complete a task, a microprocessor must often wait until all the relevant parts of a system are synchronized. The delay until the processor can proceed is a latency.

Latency is also used to describe the time it takes for an instruction to travel through a pipeline. For example, if a processor can issue and retire one instruction per cycle, an individual instruction's latency may still be 4 cycles.

Exceptions and Interrupts

Processors execute instructions sequentially and fetch subsequent instructions from the next address that the program counter points to. There are times, however, when specific software needs to get the processor's attention to handle, for example, a time-critical piece of work, or to recover from an error.

Interrupts, as their name suggests, interrupt the instruction flow in a microprocessor and cause a piece of code called an *interrupt handler* to be executed. Interrupts usually happen because a piece of hardware external to the microprocessor requires attention. Different interrupts can be caused by different parts of a computer system, and microprocessors can handle different types of interrupts in different ways. *Exceptions* and *interrupts* are generally synonymous.

Architecture

The *architecture* of a microprocessor consists of the features that are visible and accessible to the programmer who is creating software for it. A processor's architecture consists of traits that it shares with other members of its processor family.

An architecture is characterized by its instruction set, the data types its instruction set operates on, and the organization and number of registers. Implementation aspects are those features of a particular processor that aren't directly visible to the programmer; these are often features that are used to improve a processor's performance. Pipelining and superscalar design are two implementation details that software cannot affect.

Instruction Set

An architecture's *instruction set* is the collection of instructions that processors of an architecture can recognize and execute. Characteristics of an instruction set are architectural issues; for that reason, the collection of instructions that a family of processors can execute is referred to as its instruction set architecture (ISA).

All PowerPC processors share the same ISA. For example, the ISA determines that every PowerPC instruction is 32 bits long.

RISC versus CISC

The two major opposing microprocessor design philosophies are RISC (reduced instruction-set computer) and CISC (complex instruction-set computer). RISC versus CISC is an instruction-set architectural issue.

RISC processors have common traits that set them apart from their CISC cousins:

- Constant instruction length: All instructions are the same size.
- Relatively simple individual instructions: To perform complex operations, multiple RISC instructions are usually required.
- Load/store architecture: Only specific load and store instructions can read from or write to memory.

CISC processors share contrasting traits:

- Variable instruction length: Because of the complex nature of CISC instructions, they can vary greatly in size. This puts an additional performance burden on the instruction decoder in a CISC chip.
- Complex instructions: CISC instructions perform a great deal of work within a single instruction.
- Memory can be an operand: Many CISC instructions can use values in memory as operands. Since memory accesses are relatively slow, such instructions can introduce latencies that slow the system down.

One of the fundamental notions of RISC is that it is possible to execute many simple instructions more quickly than fewer complex instructions. Three basic metrics are at work here:

- CPI, cycles per instruction
- IPC, instructions per cycle
- Clock speed, the clock frequency of the processor

Performance increases as either the IPC increases or the CPI decreases.

CISC attempts to minimize the number of instructions required to perform a single task by making each instruction perform a lot of work. As a result, the IPC is reduced and the maximum clock speeds achievable with such complex microprocessor designs are limited. RISC designs emphasize higher IPC and clock speed rather than instructions per task.

One of the problems with complex instructions is that they often have internal dependencies that cannot be broken up or rescheduled via software. With the simpler RISC instruction sets, optimizing compilers can carefully schedule instructions to minimize dependencies.

Given the current state of the art in microprocessors, it appears that RISC is winning the battle. Even one of the last great holdouts—Digital, the inventor of the VAX—has admitted that RISC is the way to go and is aggressively working on siblings to the existing processors based on its Alpha architecture.

The only real CISC holdout in the desktop-computer world today is Intel, with its x86 architecture. The x86 is still managing to keep pace on the performance axis, but the current state-of-the-art Pentium is much more complex than the comparable PowerPC 601, although the two have roughly equivalent performance. The important detail to bear in mind is that the 601 is at the beginning of the upward performance curve for PowerPC, whereas Pentium is the fastest x86 processor that Intel can currently manufacture. The 604's performance leaps ahead of the Pentium; Intel's next-generation P6 processor, the successor to Pentium, is an unknown at this writing, but it will certainly compete directly with the 604.

It's important not to categorize CISC as bad and RISC as good. When CISC processors were first developed, memory was scarce and it was easier to increase software performance by throwing hardware at the problem and making the microprocessor do all the work. As technology has advanced, RISC designs and the compiler technology necessary to take advantage of them have become feasible. Many years from now, RISC may look the way CISC does from our current vantage point: RISC is a good solution to today's problems, but it isn't necesarily the end-all.

Implementation

Individual members of a family conform to an architecture specification, but each processor in a family is implemented differently. Variable implementation details include the number of functional units, a processor's clock speed, and the process used to manufacture it, as well as such other features as pipelining, superscalar design, register renaming, branch prediction, and the size of a processor's buses. Implementation details are often mistaken for architectural features. For this reason, it's important to remember the distinction between the two: *Architecture remains constant* among processors of the same family, whereas *implementation varies.*

Caches

A *cache* is a small amount of fast memory where frequently used data is stored. The purpose of a cache is to reduce the frequency with which a processor must access external memory to get a particular piece of data.

The principle of temporal *locality* is what allows caches to be useful. This principle says that software reuses both instructions and data often. Therefore, if already used data and code are kept close at hand, any speed improvement in accessing them will translate into improved processing speed, since the processor spends less time waiting for data to arrive.

Since a cache is, by definition, smaller than main memory, it has to keep track of the main memory whose data the cache contains. This bookkeeping is accomplished by *tags*. A cache tag stores the addresses of main memory that is cached.

Caches are divided into *blocks,* also known as *lines.* Depending on an individual processor's implementation, a cache block is sometimes further subdivided into *sectors.* Each cache block has its own tag, which contains the address of the memory cached within the block.

Processor caches have a characteristic known as *associativity.* A cache's associativity determines which part of it stores data found in main memory. A fully associative cache can store any part of main memory in any cache block. This is ideal, but it is also the most complex type of cache to implement.

A *direct-mapped* cache can only cache data from a particular part of main memory in a specific cache block; no other cache blocks can be used to cache data from that part of main memory.

Finally, a *set associative* cache allows several blocks—a set—to store data from a part of main memory. The number of blocks per set is specified when describing a cache's set associativity: an eight-way set associative cache, such as the one found on the PowerPC 601, has eight blocks per set. In a hypothetical 601-based system with 8MB of addressable memory space, the space would be divided among the cache's 64 sets.

In order to minimize cache contention, where the same cache block is needed to store data from different parts of memory, a particular cache set is not associated with a contiguous block of memory. Instead, the range of addressable memory is divided into increments depending on the number of sets. These chunks of memory are further subdivided into pieces depending on the size of cache blocks. In our example, the first 64 bytes at address 0 would be cached in the same set as the first 64 bytes at address 4096 (64x64=4096) and as the first 64 bytes at address 8192.

This method is used because of the principle of locality. If your software is running in the first 128 kilobytes of memory, it can use the entire cache rather than fighting over the eight blocks in the first set. See Figure 4.4.

Caches have two modes that can be changed on a per-block basis. A *write-through* cache is set up in such a way that any data written to it is written out to memory as well, thereby making sure that the two contain identical data. When writing data to a cache that is in *copyback,* or *writeback,* mode, the data is not automatically written back to main memory.

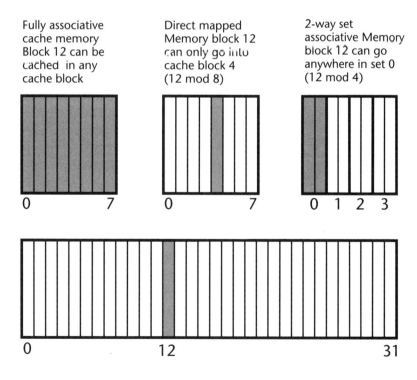

FIGURE 4.4 Cache Associativity

In addition to the different caching modes, specific parts of memory can be marked as noncacheable; any data written to such locations will not be stored within the cache. In some instances, the addresses don't refer to main memory, but are translated by a computer system's hardware and rerouted to access various input/output (I/O) devices—SCSI or Ethernet, for example. This way of accessing I/O is referred to as *memory-mapped I/O,* since memory locations are mapped, or redirected, to I/O devices. Reading back data from such addresses immediately after writing to them—which is precisely the type of interaction that a cache is supposed to optimize—should not return the written data, but should rather return the real data that's coming in from the I/O port. To avoid such a mishap, I/O addresses are marked as *noncacheable* so that writes to and reads from them do exactly what they should.

When the core of the microprocessor needs to get some data from main memory, one of two things can happen. The desired occurrence is that the address of the requested data matches the value in one of the cache tags, and the data can be transferred directly from the cache without needing to read from main memory; this is called a *cache hit*. The other possibility is that the address doesn't match any of the cache tags' contents and the data must be fetched from (slower) memory; this is known as a *cache miss*.

If a cache's blocks are subdivided into sectors, some additional information comes into play. Each sector in a cache block has an additional bit associated with it that denotes whether that sector contains valid information. This bit is called the *valid* bit. When a cache miss happens, only one sector needs to be read into the cache from RAM, minimizing the amount of bus traffic required on a miss, since a sector's worth of data is less than that of an entire block. However, this comes at the expense of increasing the ratio of cache misses.

Cache coherency is a nuisance on single-processor systems, but it can turn into a full-fledged problem in multiprocessor environments. Whenever the core of a processor writes data back to memory, it's usually in the cache. If the cache is marked as copyback, the updated information exists only in the cache but not in memory. When multiple processors share the same RAM, they must be able to know whether one processor has recently modified an address they're about to read from. If such a modification has taken place, the current data must be used rather than the stale data that resides in memory.

For this reason, *cache-coherency protocols* exist to allow multiple processors to arbitrate and determine who has the most recent version of an address's data. The protocol used in the 601 and 604 is called MESI: modified, exclusive, shared, invalid. These attributes are associated with individual cache blocks and determine the processor's behavior when it detects another processor on the system bus trying to access data in memory that it has already cached.

Caches are often referred to in connection with levels. There are Level 1 (L1) and Level 2 (L2) caches. A Level 1 cache is one that is closest to the microprocessor core. It commonly resides on the microprocessor itself—this is the case with the 6xx family of PowerPC processors—but this is not a requirement. Like a Level 1 cache, the Level 2 cache provides a buffer between a fast processor and slower main memory. A Level 2 cache is larger than an L1 cache, and it also consists of tags and cache.

The microprocessor is the key to a computer system's performance, but many different factors come into play when determining the ability of the processor to perform well. In addition to its own architecture and implementation, the design of the computer system that it's built into has a great deal to do with how well the microprocessor operates.

The following chapter is a detailed look at several members of the PowerPC family.

CHAPTER FIVE

The PowerPC Family

he PowerPC 601 processor used in the Power Macintosh 6100, 7100, and 8100, as well as in their Workgroup Server counterparts, is the first in a long line of microprocessors in the PowerPC family. All PowerPC processors, five have been announced at this writing, share common traits that make them PowerPC processors, but all five also have unique features that set them apart from each other. Some of the processors may seem to overlap and have similar features, but the announced processors have subtler distinguishing characteristics, often nontechnological ones such as price, that differentiate them sufficiently to those designing systems around them.

PowerPC is an architecture as well as the name of a family of microprocessors. PowerPC is based on IBM's POWER architecture, which was designed for high-performance UNIX workstations.

Now We're Playing with POWER

The POWER architecture, whose acronym was reverse-engineered into "performance optimized with enhanced RISC", was revealed in February 1990 when IBM shipped the first RS/6000 series workstations. IBM wasn't a complete newcomer to the workstation market, since it had shipped its ill-fated RT PC in 1986. It was based on the ROMP processor, a direct descendant of IBM's original RISC chip, the 801, which was designed in the late 1970s. When the first RS/6000s shipped, Sun Microsystems had the market more or less cornered. But within two years, IBM had garnered over 10 percent of the worldwide workstation market; clearly, it had a good product.

The original POWER architecture had standard RISC features as well as some less conventional ones. Its instructions were of fixed length, which made decoding instructions much easier, and it used a load/store architecture, where all operations are performed with data already in registers—no operations are performed directly on memory, and values must be explicitly loaded into registers or stored back into memory.

The first implementation of the POWER architecture, which IBM called RIOS at the time and now dubs Power1, had its functional units segregated and as independent as possible of each other. The separation of the functional units was largely a result of the actual implementation of Power1, where the CPU consisted of multiple-chip set. Power1's branch processor, for example, had its own register file because the bandwidth required to allow the branch processor to access registers on another chip was too great. Power1's target was maximum performance, and there was no way to fit all the functional units onto a single chip.

Power1 consisted of the following chips:

- Instruction Cache Unit: contained 8 kilobytes of instruction cache, the branch processing unit, and the instruction dispatcher
- Fixed-point unit: executed all fixed-point instructions
- Floating-point unit: executed all floating-point instructions
- Data-cache unit: two or four of these were used per system, each containing 16 kilobytes of data cache

- Storage-control unit: controlled access to memory
- I/O unit: responsible for I/O and serial ports, as well as MicroChannel cards

This division looks clean, but it has small idiosyncrasies. For example, the fixed-point unit was also responsible for address calculations used to access memory. This means that any time data has to be read from or written to memory, the fixed-point unit is responsible for calculating the correct address.

One novel aspect of the Power1 architecture was the separation of branch processing from the execution units. Traditionally, the fixed-point unit of a RISC processor is also responsible for determining whether a branch is taken and executing it. Taking a branch can have numerous side effects that reduce performance, especially in a highly pipelined environment where instructions are typically fetched well in advance of their execution. A change in program flow can cause a pipeline stall while the instructions after the taken branch are fetched.

The instruction-cache unit contains the Power1's instruction cache, the branch processor, and the instruction dispatcher. The branch processor is the key to this unit: It analyzes each branch in the instruction flow and determines, to the extent that it can, whether the branch will be taken. Depending on what the branch processor concludes, it fetches the appropriate instructions—either the one immediately after the branch or the one that the branch instruction points and that should be executed if the branch is taken—and passes them on to the correct execution unit. This technique is known as *branch folding,* since the fixed-point and floating-point units never execute branches; to them, it's a single instruction flow.

In the best case, the branch unit correctly predicts whether a branch is taken and sends the correct instructions to the instruction dispatcher. The instruction dispatcher is the next key to Power1's high performance. Since Power1's instruction-set architecture is designed to minimize dependencies between the two execution units, the instruction dispatcher can send an instruction to the fixed-point unit and the floating-point unit during the same cycle.

The big win for RS/6000 systems in the workstation market was its high floating-point performance. Surprisingly, Power1 supported only double-precision floating-point operations, which had previously been avoided where possible in favor of less accurate single-precision floating-point math for performance reasons. Since Power1's designers didn't seem to be constrained by the number of transistors used to implement the chip set, they were able to build a very high performance double-precision FPU that executed double-precision floating-point instructions as quickly as other chips were able to execute single-precision instructions.

The Power1 chip set introduced another feature that was passed on to the PowerPC family: the so-called multiply-add fused (MAF). This floating-point instruction performs a multiplication and an addition without any rounding of the intermediate result in three to four cycles; thanks to pipelining, a MAF instruction can be issued every cycle. This instruction is one of the factors that gives Power1 such high floating-point performance, since application profiling showed that floating-point software often performed this type of calculation.

Another reason that a MAF is such a useful instruction is based on the type of mathematical operation it performs: a multiply followed by an add is the basic instruction that digital signal processors (DSPs) perform. Most other processors refer to such an instruction as a MAC, short for multiply-accumulate. Although DSPs will always have a niche for specific applications where maximum performance at all times is a requirement, the presence of a fast MAC instruction can reduce the need to integrate a DSP into a computer system. Many recently announced chips, especially those going into consumer electronics devices, implement MAC instructions to reduce overall system cost by obviating the need to install a separate DSP chip.

When it first came out, Power1 provided a great deal of performance, using a complex chip set that was by no means cheap to produce. IBM realized this and designed RSC, short for RIOS single-chip, which implemented the Power1 architecture at a significantly reduced price and with less performance. RSC included a simple branch unit, a fixed- and a floating-point unit, and a unified cache, as well as memory and I/O controllers. RSC shipped in IBM's low-end RS/6000 workstations in April 1992 and was supplanted by the PowerPC 601 chip in the October 1993 release of IBM's low-end workstations.

What Makes a PowerPC a PowerPC?

PowerPC is an architecture specification: a detailed recipe that describes the way a PowerPC-compliant microprocessor behaves. PowerPC is neither a specific chip nor a kind of computer system.

IBM talks about "toasters to teraflops" when describing PowerPC's flexibility:

- At the low-end, low-cost versions of the PowerPC architecture
- At the high end, IBM's Power Parallel Systems division will use PowerPC chips in their massively parallel supercomputer designs

The family of PowerPC chips produced by Motorola and IBM will have something for everyone.

Since the design of PowerPC was a communal effort, it was important to make sure that each aspect of the PowerPC architecture specification was well documented and spelled out in great detail. The result of this effort is known as Books I through III, which together describe the PowerPC architecture. Each PowerPC processor has its own Book IV that contains that processor's implementation-specific details.

Book I defines the instruction set that a chip must be able to execute to be called a PowerPC chip. Some future PowerPC chips may not support certain instructions in hardware; if such a chip encounters an unimplemented instruction, it raises an exception that software has to handle. It's therefore possible to build more cost-effective PowerPC chips with limited functionality and have less-common instructions executed in software. An application would never notice the difference.

Book II describes the *virtual environment architecture* and details the way PowerPC interacts with storage, whether it's on-chip cache, external memory, or virtual memory. One of PowerPC's target applications is in multiprocessing systems, where multiple PowerPC chips operate in the same computer systems. For such a system to function properly, the individual chips need to know when a particular piece of memory is accessed or altered by another chip. For example, if one chip has the contents of a particular memory location in its cache and another chip changes the data in that part of memory, the first chip needs to invalidate that part of its cache so that any subsequent access to that part of memory doesn't use the wrong value.

Book II further defines how the PowerPC processors interact with memory—for example, the sequence in which burst reads and writes may be performed. The notion of storage-access ordering in the PowerPC architecture is also explained in Book II. Two specific instructions, EIEIO and

SYNC, are responsible for ensuring that certain writes to memory happen in the correct sequence. When writing to memory-mapped I/O devices, it is imperative that no optimizations are made to the order in which the writes occur; such a rearrangement could drastically affect the behavior of the intended write. EIEIO—an acronym for enforce in-order execution of I/O—is an instruction placed between write instructions to make sure that a write completes before the next one. EIEIO also affects reads for the same reason. When data is read from an I/O device, the reads must be performed in the proper sequence and shouldn't be reordered by the processor. The SYNC instruction, which behaves similarly to EIEIO, is used in instances where the memory write is not to a memory-mapped I/O device.

Book III defines the PowerPC *operating environment architecture*. It defines a PowerPC processor's lowest-level operations and their results. The state of the processor, interrupt handling, memory protection, and address translation are all defined in this volume.

Finally, Book IV describes individual chip implementations. There is no generic book four. The 601 and 603 user manuals, for example, are sanitized versions of the Book IVs for those chips. A Book IV contains chip-specific implementation details: instructions supported in hardware, instruction timings, cache implementation, and suchlike. Assembly language programmers (yes, there are RISC assembly-language programmers), compiler writers, and hardware designers, and those trying to divine the greater meaning of code generated by a compiler, are the primary beneficiaries of a user manual.

The Abstract PowerPC

Despite the great detail that Books I through IV go into, the basic features of PowerPC are straightforward, as shown in Figure 5.1.

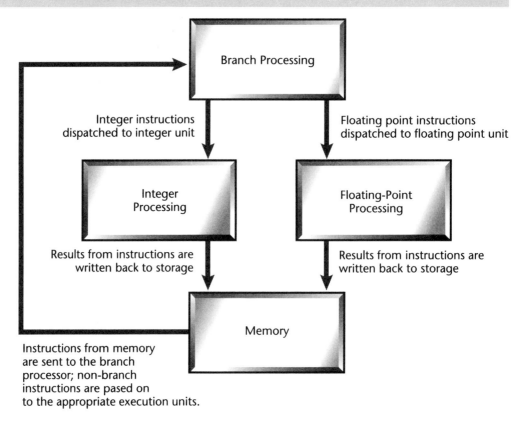

FIGURE 5.1
The Abstract PowerPC

The PowerPC architecture is derived from Power1. The idea behind PowerPC was to make single-chip, high-volume, low-cost implementation feasible and also to remove any limitations to scalability of the architecture.

Similar to Power1, each PowerPC chip has a branch processor that deals with branches and the resulting instruction dispatch. Each PowerPC chip also has a fixed-point execution unit, which performs all fixed-point calculations. On the PowerPC 601, the fixed-point unit is also responsible for performing address calculations and executing loads and stores for the general-purpose registers. In contrast to Power1, PowerPC's floating-point support includes single-precision operations—for reasons of chip real estate, it makes sense to support single precision in a single-chip processor. In addition, typical floating-point-intensive applications, such as renderers or other graphics apps, don't need the additional precision provided by doubles. Finally, double-precision operands are twice as large, 64 bits versus 32 bits for singles, so using single-precision operands can actually result in memory savings for floating-point-intensive software.

One of the main differences for the Macintosh universe is that the floating-point part of the PowerPC architecture doesn't support the 80-bit extended floating-point format, which is the native format for 68k floating-point hardware as well as for Apple's Standard Apple Numerics Environment, the floating-point system software available on every Mac. As a result, those applications that depend on extended floating-point operations need to change to use either single- or double-precision floating-point operations.

The individual execution units all interact with memory through the cache and the chip's bus. Caches are implementation-specific, so there's no specification on how a PowerPC cache is designed or how it should behave. The differences in cache implementation between the 601 and 603 alone are quite drastic.

Compared to Power1, PowerPC implements changes designed to facilitate future, more superscalar implementations. This approach exemplifies the so-called brainiac versus the speed-demon approach. Today's RISC processors can be grouped loosely into two camps:

- Those that achieve increased performance by raising the clock frequency
- Those that increase performance by adding more parallelism and more execution units within the chip

Alpha and MIPS clearly are speed demons with their high frequencies. PowerPC is a brainiac, since it uses multiple independent execution units to perform its tasks. It also executes more complex calculations, such as multiply-add fused.

The differences between Power1 and PowerPC are subtle to the casual observer, but all the changes share common goals. The goals of the PowerPC architecture are as follows:

- Allow a broad range of implementations, from low-end embedded chips to high-performance superscalar versions
- Support multiprocessing (multiple PowerPC processors running in the same system)
- Remove limitations that would hinder superscalar implementations
- Define 32- and 64-bit operating environments

These goals have been achieved. The PowerPC chips are detailed in the following sections.

The PowerPC 601

The 601 is the first member of the PowerPC family. It has all the standard features of a 32-bit PowerPC processor, but it also has characteristics that set it apart from its successors. The main goals of the 601 design were as follows:

- Fast time to market
- Serve as a bridge between the POWER and PowerPC
- Provide high performance

The 601 was designed for use in desktop computers, and although IBM has shipped a laptop based on the 601, this doesn't make the 601 especially suitable for use in hardware that stays away from power sockets for prolonged periods of time. Apple's backward-compatibility solution for existing Mac software is to provide a 68k emulator, but IBM had an existing installed base of POWER-based workstations whose software couldn't be converted immediately. For this reason, the 601 supports not only the PowerPC instruction set, but it also implements the Power1 instruction set and can execute Power1 code. The 601 will be the only processor with this degree of backward compatibility in hardware. Subsequent PowerPC chips support only the PowerPC instruction set.

The 601's design is based on work that started out being called RSC+. At the time that the PowerPC alliance came into being, IBM's designers had a follow-on chip to the original RSC chip in the works, and its design was used as the basis for the 601. This approach had the benefits of not having to start from scratch, making quick time-to-market possible.

The 601 is a hybrid chip from an instruction-set perspective, since it executes the Power1 instruction set as well as the PowerPC instruction set, albeit both with minor exceptions, none of which will ever be noticeable to users or high-level-language programmers.

The 601's bus design was based on the work that Motorola did for the 88110, the chip that at one point was going to be the basis for the RISC Macintosh. Apple already had an investment in logic-board and support-chip designs that assumed an 88110 bus; the 601's bus is similar enough to the 88110's that only minor modifications had to be made to existing designs to support the 601.

All versions of the 601 are built by IBM at its manufacturing facilities. Motorola will begin PowerPC production with the 603; customers purchasing 601s from Motorola receive IBM-fabricated parts. Since fast time-to-market was one of the primary goals of the 601 design, the companies agreed to let IBM be the sole manufacturer of the 601 using an IBM-only 0.6μ process. All subsequent processors are built by both IBM and Motorola. Except for the sub-100MHz 601, the two manufacturers use exactly the same fabrication processes, so an IBM-built PowerPC chip has no inherent process-related benefit over one made by Motorola.

Basic Features

Like all 32-bit PowerPC processors, the 601 has 32 general purpose registers (GPRs), each of which is 32 bits wide. The 601 also has 32 floating-point registers (FPRs), each of which is 64 bits wide, the size of a double-precision floating-point number.

The 601's connection to the outside is via a 64-bit-wide data bus and a 32-bit-wide address bus. It can issue up to three instructions per cycle: a branch, a fixed-point instruction, and a floating-point instruction.

Cache, Bus, Memory: The 601 has a single 32 kilobyte unified cache. There are no separate caches for data and for instructions. The cache is eight-way set-associative, which means that data from a particular location in memory can be stored in one of eight cache blocks in a set. Each cache block in the 601 is 64 bytes (not bits) in size. The cache has eight sets of 64 blocks, making for a total of 512 blocks in the cache. On the 601, each cache block is further subdivided into two sectors of 32 bytes each; a sector's worth of data can be transferred during a single 4-beat burst transaction on the bus.

The 601's bus is the standard PowerPC 60x bus and is compatible with the buses of the other members of the PowerPC 60x family. This bus is a derivation of the Motorola 88110 bus; it supports so-called *split transactions*, where the address bus and data bus are performing two different transactions simultaneously.

Multiprocessing Support: The 601 is also designed with multiprocessing in mind. It supports the MESI protocol, which allows a cache block to be declared modified, exclusive, shared, or invalid. These states are important when multiple processors share the same memory space. If one processor has a piece of memory cached and has modified it, this fact needs to be communicated to the other processors in the system so that they can take the necessary precautions to make sure they're working on the most recent data.

Finally, the 601 can operate at an integer multiple of its bus frequency. In the case of the Power Macs, the 601's speed is two times that of the bus speed. The Power Macintosh 6100/60 runs its 601 at 60MHz and its system bus at 30MHz.

When the 601 was first released, speeds of 50MHz and 66MHz were announced. At the May 1992 Apple Worldwide Developers' Conference, an 80MHz technology demo was shown, only to have the PowerPC alliance announce the availability of an 80MHz 601 a few months later. In late March 1994, the alliance announced a 100MHz 601. The original 601 was manufactured using a 0.65µ process, and the faster 100MHz 601 is manufactured using a newer 0.5µ process. Consequently, the die size of the 100MHz 601 is smaller than that of the first generation 601s. No changes were made to the 601's design per se; the size difference is exclusively a matter of the new process.

Execution Units

The PowerPC 601 contains three main execution units: the branch-processing unit (BPU), the integer unit (IU), and the floating-point unit (FPU). In addition, the 601 has a memory-management unit (MMU) and a bus-interface unit (BIU). See Figure 5.2.

As a result of the way the BPU, IU, and FPU work together, the 601 supports out-of-order dispatch. This means that instructions can be issued to execution units even if preceding instructions for another execution unit are still

FIGURE 5.2
The PowerPC 601

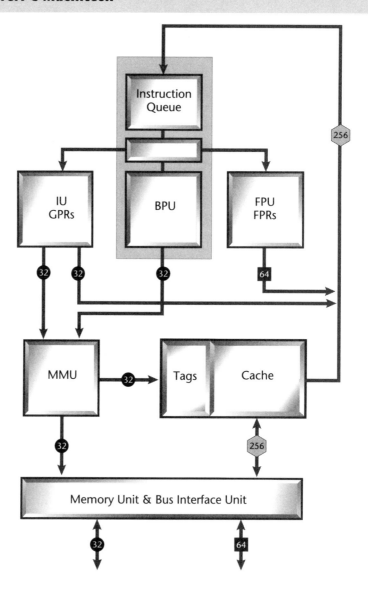

waiting to be issued. Out-of-order execution by itself would be a problem, since it's important that instructions complete in the order that they appear in software; the 601 ensures in-order completion of instructions when necessary. Branch instructions, for example, are executed as early as possible.

The Branch-Processing Unit: The 601's branch-processing unit fetches instructions to be executed from the instruction cache, decodes them, and issues them to either the fixed-point or the floating-point unit, whichever is appropriate. Branch instructions are processed within the BPU. The 601 tries to predict whether a branch will be taken, and fetches subsequent instructions depending on its prediction.

The 601's branch-prediction scheme, however, is static and not as sophisticated as that on higher-end RISC processors such as 604. The static prediction scheme assumes that branches backward—ones that point to an address in memory prior to that of the current program counter—will be taken. Since much software consists of often reexecuted code, the same characteristic that makes caches worthwhile, this assumption is frequently correct. Branches forward—those that jump beyond the instruction immediately after the branch—are assumed by the 601's BPU to be not taken. In instances where the branch prediction is false, a time penalty is incurred while the BPU fetches the correct instruction. This default prediction can be changed by the compiler by changing a bit in the instruction; if that particular bit in a branch instruction is set, the BPU assumes the opposite of its default about the direction the branch will take.

When the BPU guesses right about the branch, it performs what is known as branch folding. In most other microprocessors, the fixed-point unit is in charge of executing branch instructions, so branches occupy space in the pipeline and take time to be executed. By contrast, the 601's IU and FPU *both* see continuous streams of instructions without interruption by branches; the BPU removes the branches from the instruction stream (it folds them away). This also means that branches are executed in zero cycles, since they have no effect on the performance of either the IU or the FPU.

As a result of the BPU's branch folding, the IU and FPU can run without frequent branch-based interruptions. This is a core factor in the 601's high performance.

The Integer Unit: The 601's IU is responsible for executing all fixed-point, also known as integer, instructions. In addition to performing addition, subtraction, multiplication, and division on data in one of the 32 general-purpose registers, the IU is also responsible for any address calculation required for any load or store operations, regardless of whether they are integer or floating-point loads and stores. The IU also performs fast comparison between two operands and forwards the result to the BPU, which uses this information to efficiently process subsequent branch instructions that depend on the outcome of the comparison. The IU is also responsible for performing all loads and stores to the GPRs. Unlike later members of the PowerPC family, the IU has a great deal more to do than just integer math.

The Floating-Point Unit: The 601's floating-point unit supports operations using either 32-bit single-precision or 64-bit double-precision floating-point values. As with the IU, the FPU is responsible for all loads and stores of the 32 floating-point registers. The FPU is compliant with the IEEE-754 standard for single- and double-precision floating-point operations.

In addition to support for the standard operations, the FPU contains hardware to perform single-precision multiply-add fused (MAF) as well as double-precision MAF. This instruction is executed more quickly than individual multiply and add instructions would be if they were issued in sequence. The MAF instruction contributes significantly to the 601's floating-point performance, provided that compilers generate code that takes advantage of it.

The 601's floating-point unit also supports a mode where exceptions caused by floating-point instructions aren't raised immediately, but rather a few instructions later, the pipeline permitting. This allows floating-point code to execute faster, but at the expense of not being able to catch floating-point exceptions immediately.

The Bottom Line

The 601 is at the heart of the first generation of PowerPC-based machines from Apple, IBM, and others. Its success will make or break the future of the PowerPC family. At this writing, a vigorous battle was under way with Intel, which announced the availability of 90 and 100MHz Pentium chips one week before the introduction of the Power Macs. IBM has been shipping 601-based UNIX workstations since October 1993, which met with favorable reviews. The first weeks of the Power Macs' availability indicate that the 601 is a success in Apple's machines as well.

The 100MHz 601 will offer even better performance than was originally anticipated from the 601, and at much lower cost; smaller die size means lower chip prices. This high-speed version is also helpful for IBM's midrange and low-end workstation business, since it extends the time period for migration from Power1 to PowerPC by providing increasingly higher performance.

At this writing, Apple had not announced plans to use the 100MHz 601 in forthcoming Power Macs. Historically, however, Apple has often provided so-called speed-bumped new versions of existing Macs. The Quadra 900 to 950 transition was a speed boost from 25MHz to 33MHz; the Centris 610 running at 20MHz later became the Quadra 610 running at 25. And since the Power Macs contain the processor frequency in the product designation, it would be easy to change just the frequency without confusing anyone about the individual Mac's features set.

The 601's statistics are shown in Table 5.1.

Table 5.1 601 Statistics

Speed	80MHz	100MHz
SPECint92	85	110
SPECfp92	105	130
Voltage	3.6V	2.5V
Power (max)	8 Watts	4 Watts
Size	120mm^2	74mm^2
Process	0.6μ	0.5μ
Transistors	2.8 million	2.8 million

The PowerPC 603

The 603 is the second member of the PowerPC family to be announced and produced. First silicon for the 603 was announced in October 1993, and high-volume production was scheduled for summer 1994. The 603 has two firsts in the PowerPC family:

- The 603 is the first PowerPC chip to implement the PowerPC architecture and no other; the 601's POWER backward compatibility is not supported.
- The 603 is the first PowerPC chip to be produced by both IBM and Motorola. For the first time, the two manufacturing members of the alliance will be competing against each other on the merchant market with chips of their own fabrication.

The 603 is designed as a low-cost, low-power processor. It is perfect for use in laptops, but also for low-cost, high-volume desktop machines. At this writing, Apple had announced that PowerBooks based on the PowerPC 603 will be available in the first half of 1995. Given also the success of LC-class 68k Macs, it also stands to reason that 603-based desktop machines are probably being developed as well.

Even the lower-priced 100MHz 601 will cost more than the 603. And when it comes to extremely low-priced desktop systems that must compete with x86 machines, every dollar of materials cost is important.

Basic Features

The 603 is a completely new design, not based on any previous processor designs from IBM or Motorola. Consequently, it has several design differences when compared to the 601. Where the 601 used the integer- and floating-point units for loads and stores, the 603 has a dedicated load/store unit that handles the mechanics of moving data between registers and memory, including calculation of addresses. The 603's integer unit is free to concentrate completely on performing the duties of an integer unit. See Figure 5.3.

Cache, Memory, Bus: The second major difference is that the 603 has separate caches for instructions and data. Each cache is 8 kilobytes of two-way set-associative cache. The cache-block size is 32 bytes. The 603 can be integrated into a system with either 32-bit or a 64-bit data-bus width. The former allows for more inexpensive designs at the cost of performance. With a 64-bit-wide bus, the 603 supports single-beat transactions of 1 to 8 bytes, as well as 4-beat, 32-byte bursts. With a 32-bit data bus, the 603 allows single-beat, 1 to 4 byte transactions, as well as 2- and 8-beat bursts.

The 603 supports clock-speed to bus-speed ratios of 1:1, 2:1, 3:1, and 4:1. A 66MHz 603 supports bus speeds of 66MHz, 33MHz, 22MHz, and 16.6MHz. The 603's bus is the standard PowerPC 60x bus and is compatible with the buses of other PowerPC 60x chips.

Multiprocessing Support: The 603 was not designed to operate in a multiprocessor environment; it doesn't contain

FIGURE 5.3
The PowerPC 603

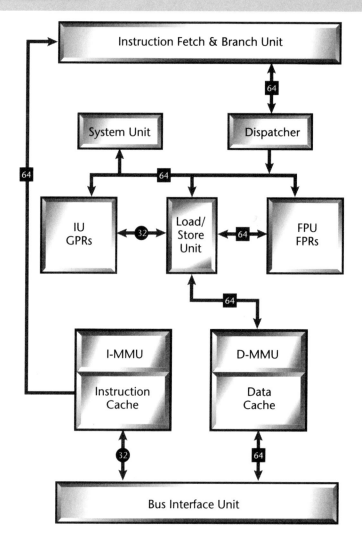

support for the full MESI protocol as the 601 and 604 do. Instead, it implements a subset that is sufficient for the 603 to coexist on a bus with other bus masters such as direct memory access (DMA)-capable devices. Each cache block in the 603 can have either exclusive, modified, or invalid attributes; the shared attribute necessary for multiprocessing is not available.

Power Management: The next main change from the 601 is the 603's power-saving modes, of which there are four. Full-power is, as its name suggests, the most energy-inefficient mode, but it also provides the most performance. Even when in full-power mode, the 603 consumes less than 3 Watts of power when running at 80MHz. The full-power mode is the default mode for the 603, but it allows a dynamic power-management mode that selectively disables functional units that are idle, without any part of the system being any wiser. When a functional unit disabled in this fashion is needed again, there is no lag or any other penalty to pay for having enabled the dynamic mode.

Doze mode is the first major power-saver mode. It disables all functional units on the 603 except for the unit that maintains the clock on the bus, the part of the chip that tracks data moving across the external data bus, and the on-chip timers. Even when running at 80MHz, the 603 consumes less than 0.5 Watts in doze mode. A downside of this mode is that it takes a few processor cycles to bring the 603 back up and into a fully functional state. This lag, however, is rarely problematic for software.

Nap mode is a step beyond doze mode. In nap mode, the 603 stops tracking what's happening on the data bus; only the on-chip timers are still running. The power consumption is less than half of doze mode's. Compared to consumption at full-power mode, these last modes approach the point of diminishing returns, as the 603's power use becomes infinitesimally small.

Finally, sleep mode is the most power-frugal of them all. It disables all the 603's internal functional units, and the computer system can turn off the external timers as well. Waking a sleeping 603 is considerably more work than resuscitating it from any of the other modes. On the other hand, the chip uses hardly any power in this mode, so if you know that the machine will be dormant for a while, this mode is an excellent alternative.

Execution Units

The PowerPC 603 has five execution units and is able to issue a total of three instructions per cycle.

Branch Processing Unit (BPU): The 603 has a branch-processing unit as well. It performs the same duties as that of the 601, with some additions thrown in. The 603's BPU constitutes a superset of the 601's. Like the 601's, it provides the ability to execute and fold branches so that the integer and floating-point units never have to contend with processing a branch. The BPU also performs branch prediction, using the same static prediction scheme that the 601 does. The difference in the 603's BPU lies in its ability to calculate branch addresses by itself rather than relying on the integer unit (IU) to calculate the addresses needed by the BPU. This further unburdens the IU from maintenance tasks.

Integer Unit (IU): The integer unit is responsible for processing all integer instructions. Most integer instructions on the 603 take a single cycle to execute. Unlike the 601, the register file for the GPRs is not an integral part of the integer unit. The general-purpose register file contains 32 32-bit GPRs, as the PowerPC specification requires. In addition, the GPR file also contains five rename buffers.

Register renaming is a performance-enhancement technique used to keep the pipeline flowing. One of the most common instances of resource contention within a microprocessor occurs when multiple instructions want to write their data to the same register. The hardware of the renaming scheme examines the instructions in the pipeline to make sure that there are no interdependencies where one instruction relies on the value that the other placed in the conflicting register. Once it's determined that no conflict exists, one of the values is written to a rename register, so titled because the register has been temporarily renamed to that of another. The data in the rename register is written into the real register once the conflict has gone away.

Floating-Point Unit (FPU): The 603's floating-point unit supports the same basic functionality as the 601's. It's IEEE-754 compliant with regard to single- and double-precision arithmetic. Also like the 601, the 603's FPU supports the multiply-add fused (MAF) instruction. Like the 603's IU, the 603's FPU also has rename registers at its disposal to handle register conflicts without having to stall the pipeline.

The 603 had an additional floating-point mode, called NI, for non-IEEE. Informally, this mode is referred to as sleaze mode. The IEEE 754 specification is explicit about how to handle denormalized floating-point numbers. Numbers with a zero exponent field and a zero fraction are defined as being zero. Numbers that have a zero exponent field and a nonzero fraction part are known as denormal numbers and are defined carefully by the specification as having a specific value; with sleaze mode enabled on the 603, such a number is quietly treated as zero. The 603's sleaze mode is useful for software that requires the highest performance but not the utmost accuracy.

Load/Store Unit (LSU): The PowerPC 603 is the first PowerPC processor with a separate load/store unit. This execution unit performs the work necessary to move data between the register files—both the GPR and FPR files—and memory, including cache. The LSU doesn't rely on the integer unit to calculate addresses; it performs these calculations internally.

System-Register Unit (SRU): The system-register unit executes the various system-register instructions that don't fit in with any of the other execution units. The SRU performs logical operations on the condition registers to determine their status, and it moves data to and from special-purpose registers.

Completion Unit (CU): The 603's completion unit isn't a functional unit like the others. It is more like the MMU or caches, since it does not execute instructions directly. The CU guarantees that integer and floating-point operations complete in the order that they appear in the incoming instruction stream. This feature is important when out-of-order execution takes place, since instructions don't necessarily execute in the same order as they are found in memory. The completion unit makes sure that the results of operations happen in the sequence that the executing software expects them to.

The Bottom Line

The 603's smaller split caches will cause the performance of 603-based Macs to be less than that of 601-based ones, since the 68k emulator is a major beneficiary of the 601's large unified cache. Since the 68k instructions are treated as data by the 603, there will be more cache reloads on a 603-based Mac. It is certainly possible to add a Level 2 cache to a 603-based Mac to compensate for the smaller cache, but this would raise the price of such a system significantly as well as lose much of the benefit of lower power consumption. Also, if the 603 is implemented using its 32-bit-wide data-bus option, available bandwidth between the chip and the rest of the system is decreased, leading to further performance degradation.

Despite these performance-related issues, the 603 looks to be a promising chip for mobile systems as well as for low-cost, lower-power desktop machines.

The 603's statistics are shown in Table 5.2.

Table 5.2 603 Statistics

Speed	66MHz	80MHz
SPECint92	60 (estimated)	70 (estimated)
SPECfp92	75 (estimated)	85 (estimated)
Voltage	3.3V	3.3V
Power (Macs)	2.5 Watts	3 Watts
Size	85mm^2	85mm^2
Process	0.5μ	0.5μ
Transistors	1.6 million	1.6 million

The PowerPC 604

The PowerPC 604 was announced in April 1994. It's the first pure PowerPC processor designed for the desktop. Its performance is suitable for midrange to high-end desktop machines as well as servers. Like the 603, the 604 doesn't support any of the original POWER instructions that the 601 docs.

Basic Features

The 604 is easily the most complex of the PowerPC processors known today. It has many similarities with the 603's microarchitecture, including a dispatch bus, a completion bus, and a completion unit that tracks instructions from dispatch through execution. This ensures that they are completed in the order they appear in the instruction flow, regardless of whether they were executed out of order. See Figure 5.4.

Cache, Memory, Bus: The 604 has two 16-kilobyte, four-way set-associative caches, one each for data and instructions. The 604's cache-block size is 32 bytes. The 604's interface to the outside world is through a 32-bit address bus and a 64-bit data bus. There is no provision for a 32-bit data

FIGURE 5.4

The PowerPC 604

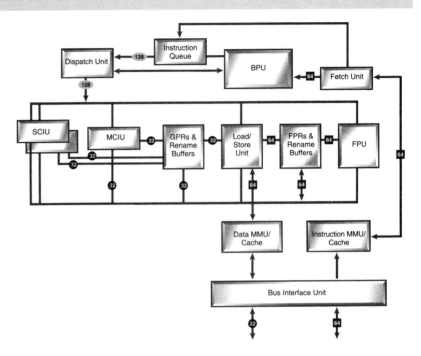

bus. Single-beat transactions of 1 to 4 bytes are supported by the 604, as are 4-beat-burst reads and writes.

The 100MHz 604's bus supports speeds up to 66MHz and can be run at ratios of 1:1, 1.5:1, 2:1, and 3:1 to the processor speed. This means that the 100MHz 604 supports bus speeds of 66MHz, 50MHz, and 33MHz. The 604's bus is the standard PowerPC 60x bus, and it is compatible with the buses of the other members of the PowerPC 60x family.

Multiprocessing Support: Like the 601, the PowerPC 604 supports the MESI cache-coherency protocol, which allows cache blocks to be designated as modified, exclusive, shared, or invalid. This allows multiple 604 processors, or multiple bus masters such as devices that support direct memory access, to share the same main memory and allow all parties to communicate with each other about whether a particular part of memory is cached or in use.

Power Management: The 604 contains support for a nap mode, during which all internal processing and bus operations are suspended. Nap mode is enabled through a software instruction; when a 604 is napping, its internal timers are still active, but it does not pay attention to data traveling over the external data bus.

Execution Units

The 604 has the largest number of execution units of any announced PowerPC processor. Most of them are familiar from the 603: a floating-point unit, a load/store unit, a branch-processing unit, a dispatch unit, and a completion unit. Unlike any previous PowerPC processor, the 604 has three integer units, two of which are identical.

Integer Units: The 604's general-purpose register file contains 32 32-bit GPRs, and it also has 12 rename registers. Like the 603, the 604 supports register renaming to avoid stalling.

The 604 has two single-cycle integer units (SCIUs). They execute only those integer instructions that can complete within a single cycle: additions, comparisons, and logical operations, as well as rotate and shift operations.

The single multicycle integer unit (MCIU) in the 604 performs the more complicated integer operations: multiplication and division. The MCIU is not a superset of the SCIUs; it does not support any of the arithmetic operations that they do.

Floating-Point Unit: Like the FPUs in the 601 and the 603, the 604's FPU is compliant with the IEEE-754 standard for single- and double-precision floating-point arithmetic. Although the 604's FPU is faster than those of the other two PowerPC processors, it behaves exactly the same way.

The 604's floating-point register file contains the standard complements of 32 64-bit FPRs; in addition, it also has eight rename registers to help avoid stalls.

Load/Store Unit: As in the 603, the 604's load/store unit (LSU) is responsible for moving data between memory/cache and the register files. It performs all the necessary address calculations required to determine the source or target address of a transaction.

The LSU on the 604 allows speculative load operations that precede already pending store operations. It also performs the work necessary to resolve dependencies between the data in the pending store operation and the speculative load if the two transactions go to the same addresses in memory.

Decode/Dispatch Unit: The 604's decode/dispatch unit (DDU) works closely with the branch-processing unit to keep instructions flowing to the execution units as quickly as possible. In the 601 and 603, the branch-prediction logic is static: The BPUs in these processors always predict branches the same way. Unlike the BPU on the 601 or 603, the BPU in the 604 isn't responsible for branch prediction. Branch prediction on the 604 is dynamic and performed by the DDU. The DDU contains logic that evaluates the likelihood of the direction to which a branch will resolve, updated every time a branch is executed.

In the 604, the first time a branch is encountered, the DDU takes note in its branch-history table (BHT) and remembers which way the branch went. Every time this branch is executed again, the DDU updates the information in the BHT, depending on whether the branch was taken. Each entry in the BHT can have one of four values: strongly-taken, taken, not-taken, and strongly-not-taken. Each time a branch is executed, its entry in the BHT is incremented for each branch taken, decremented for each branch not taken. This way, when a branch is first encountered and it's taken twice in a row, the DDU predicts the next instance of that branch as strongly-taken. This dynamic branch prediction is more accurate in the long run than the static prediction of the 601 and 603, since it's based on past history rather than fixed assumptions.

Branch-Processing Unit: The 604's BPU is different than those of the 601 or the 603. Unlike its predecessors, the 604's BPU doesn't perform any branch folding. Branch instructions are issued to the BPU by the 604's dispatch unit just like a floating-point instruction would be issued to the FPU. The BPU in the 604 processes the branch based on the prediction of the DDU whether the branch will be taken.

The Bottom Line

The 604 is a big leap in performance beyond the 601. At the same clock frequency, the 604 outperforms the 601 by 60 percent in integer operations and by approximately 27 percent in floating-point operations. The 604 will be used in midrange and high-end systems, where price and power consumption are less of an issue. By comparison, the 603 will be used in extremely low-cost desktop systems as well as portable machines; the 601 will be used in the low-cost to midrange systems, just like today's crop of Power Macs.

The 604's statistics are shown in Table 5.3.

Table 5.3 604 Statistics

Speed	100MHz
SPECint92	160 (estimated)
SPECfp92	165 (estimated)
Voltage	3.3V
Power	< 10 Watts
Size	196mm^2
Process	0.5μ
Transistors	3.6 million

The PowerPC 403GA

IBM's PowerPC 403GA is a so-called embedded controller: It's designed to be used in dedicated hardware such as laser printers and television set-top boxes. The 403GA will never be used in desktop systems such as Macs. The 403GA is included here to illustrate that the PowerPC architecture is more far-reaching than simply a line of processors for mainstream desktop systems.

The PowerPC 4xx series is designed and produced by IBM alone. The PowerPC alliance allows its members to build their own PowerPC variants, as long as these variants conform to the PowerPC architecture specification. IBM's 4xx series will be a line of embedded microprocessors with feature sets reflecting their intended use. The 403GA is the first member of that line, and it is a general-purpose microcontroller with features that make it usable in a variety of situations.

Motorola is known to be working on its own family of embedded processors, the 5xx family, but at this writing had not made any public announcements about features or availability of specific 5xx processors. Ford Motor Company, however, is using a Motorola-supplied embedded PowerPC processor for at least its next-generation transmission computers.

Features

The 403GA is a 32-bit PowerPC processor. It contains a general-purpose register file with 32 32-bit GPRs, and an execution unit that performs one-cycle integer arithmetic, shift, rotate, and logical operations. Its branch processor performs branch folding as well as static branch prediction. The 403GA has a 2-kilobyte instruction cache and a 1-kilobyte data cache; both are two-way set-associative and have a cache-block size of 16 bytes. See Figure 5.5.

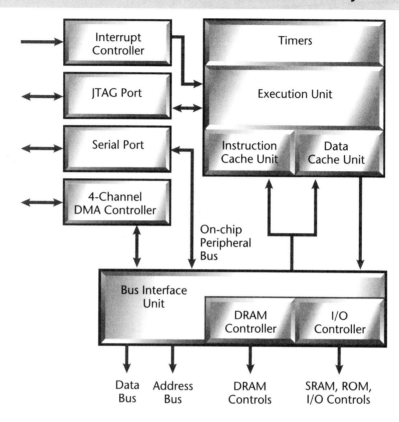

FIGURE 5.5
The IBM PowerPC 403GA

The 403GA does not contain a floating-point unit; the additional cost of a FPU would make the 403GA unnecessarily expensive. In the embedded controller market, where devices containing the controllers are made in much larger volumes than, for example, personal computers, the sensitivity of additional cents per controller is very high. The less the controller costs, the likelier it is to be adopted. In addition, embedded microcontrollers don't traditionally need to perform many floating-point calculations anyway.

The Bottom Line

The 403GA is proof of IBM's toasters-to-teraflops claim about the PowerPC. The 403GA is clearly on the toaster side of the axis where low cost-per-unit is paramount. Although the 403GA isn't at all relevant to the Macintosh or the personal-computer market, it is an indicator of the flexibility of the architecture. In the embedded market, the adoption by a single large-volume customer can make the difference between a successful and a mediocre product. Intel's i960 processors were chosen by Hewlett-Packard for its LaserJet 4 family of printers; this decision catapulted the i960 into the lead as the highest-volume RISC processor. With the 6xx series on the desktop, and the 4xx and 5xx series in the embedded controller market, the PowerPC looks to have a good shot of attaining that designation in the near future. The 403GA is statistics are shown in Table 5.4.

Table 5.4 403GA Statistics

Speed	25MHz
Voltage	3.3V
Power	1.2 Watts
Size	$39.4mm^2$
Process	0.5μ
Transistors	585,000

The PowerPC 620

At this writing, first silicon for the PowerPC 620 was unannounced. Its designation is known from the original PowerPC alliance announcement, and a few technical details are also available, but no in-depth information about this chip is publicly available.

The 620 will be the first 64-bit PowerPC processor. The 601, 603, and 604 are all 32-bit processors; their GPRs are 32 bits wide, and the internal workings of their integer units are based on a 32-bit unit of data. The 620 will have 64-bit-wide

GPRs and the additional instructions to support operations on these larger integer numbers.

The 620 is also expected to be by far the fastest PowerPC processor; it will be a true high-end chip. The 604 is a good hint that future performance improvements may happen as a result of additional execution units. During the 620's design, when a decision had to be made between cost and performance, performance was chosen. The 620 will not be an inexpensive chip, nor will systems based on it be inexpensive.

Typical applications for the 620 will be high-performance servers. IBM has already alluded that it plans to put a 620 into future AS/400 minicomputers, as well as in parallel supercomputers. IBM has a separate division, Power Parallel Systems, that has already built parallel computers based on IBM's Power2 chip set; it is expected to build a lower-cost system around the 620 when the chip becomes available.

It's unclear how soon after its introduction 620-based Mac systems would be available. The 620 has the ability to run in 32-bit mode, so it will provide backward compatibility with today's PowerPC processors and their software. However, a 620-based Mac, if it happens, will probably be the most expensive Mac ever, and since the personal-computer market is so competitive on price, a 620-based Mac might not even make sense to build since so few would want to buy it. On the other hand, recent market studies show that the primary reason for the strong sales of the Quadra 660AV and Quadra 840AV had little to do with the machines' AV capabilities—buyers wanted the fastest Macs available at the time, and the AV Quadras were the ones.

However, this is all speculation. The 620's first silicon is expected to be announced before the end of 1994.

The PowerPC family's breadth and depth has already been made evident by its first four members. The three first 6xx series PowerPC processors were delivered on schedule, proof that the Apple/IBM/Motorola alliance is working.

CHAPTER SIX

Emulators on the Power Macintosh

he promise of the Power Macintosh is that it will offer unprecedented performance in a personal computer, performance that until now has been seen only in high-end computers such as engineering workstations. These fast workstations never succeeded outside their own niche in the computer market partially because they were expensive, but mainly because hardly any of the mainstream productivity applications, such as word processors and spreadsheets, were available for the OSs and processor architectures in the workstation world.

The Power Macs won't have to face this problem, since existing Macintosh software runs on the new Macs, even though the PowerPC family of microprocessors cannot execute 68k code directly. This compatibility with existing 68k-based software is courtesy of an *emulator*—software in the Power Macs' ROM that interprets 68k code and performs the 68k code's work on the PowerPC chip. The only drawback of the emulator is speed. Since it takes more time to perform the interpretation, performance of emulated software is roughly the same as if the software were running on a high-end 68030- or low-end 68040-based Mac and not nearly at native speeds.

One other emulator is available for the Power Macs, although it's neither part of the ROM nor an Apple product. Insignia Solutions has developed a software package called SoftWindows that runs on Power Macs. SoftWindows emulates an 80286-based PC running MS-DOS and Windows 3.1. The performance of this emulator is roughly equivalent to a high-end 80386 or low-end i486 PC.

The emulation offered on the Power Macs is investment insurance and models, the migration to native software. It will take some time before most Mac software runs native on PowerPC, and some 68k software, especially if it's old, may never run native on a Power Mac.

With these two emulation solutions—one built into every Power Mac and the other available as an option—the new PowerPC-based Macs are easily the most versatile and compatible personal computers available today.

Emulation Works

Using emulation as a transition strategy isn't a new idea. IBM used emulation successfully to help its mainframe customers make the transition from its 1401 mainframes to its 7094 series. Later, it made another transition from the 7094 to the IBM 360 series.

IBM, however, wasn't the only company to successfully use emulation to make a transition survivable for its customers. DEC included PDP-11 emulation in its VAX minicomputers for the same reasons: The installed base of software was a valuable investment for its customers, and making it obsolete would only serve to alienate its customers.

The 68LC040 Emulator

From the user's perspective, the Power Macs' 68LC040 emulator is integrated seamlessly into the Power Macintosh environment. There's no way to tell whether emulated code or native code is running. And emulated software never has any idea that it's not running on a 68k-based Mac. The only clue that software is running in emulation is performance: Emulated code runs considerably slower on the Power Macs than native PowerPC code.

Insurance

The 68LC040 emulator in every Power Mac's ROM is insurance for users, developers, and Apple alike. Without it, there would be no smooth transition from 68k Macintosh to PowerPC Macintosh—it would be as if the two were completely different systems. With the 68k emulator, however, existing 68k-based Mac software can run on the new Macs, albeit with less performance than native apps. Still, users' existing investment in Mac software isn't made suddenly worthless. In fact, most developers of popular Macintosh apps will be offering inexpensive upgrades from 68k to native PowerPC software, unfortunately, at this writing, a few Macintosh software vendors are trying to make this migration into a profit center.

But developers and users aren't the only beneficiaries of the emulator; Apple benefits as well. Not all of the system software for the new Power Macs is completely native. Had Apple decided to wait until the entire operating system was native, the Power Macs would never have shipped when they did. The system software running on the Power Macs is as dependent on the emulator's compatibility and reliability as third-party software is.

A side effect of having parts of the OS remain as 68k code is a high degree of compatibility, since some of the code used in the Power Macs is, in fact, identical to code in 68k Macs. Consequently, existing software that works on 68k-based Macs is likely to work with the same 68k code running in emulation on the PowerPC-based Macs.

68LC040 Emulation

The emulator in the Power Macs acts like a 68LC040 processor in the processor's so-called user mode. The 68k processors also have a so-called supervisor mode, which allows the execution of special instructions that control, for example, the MMU and on-chip cache; these instructions cannot be executed in user mode. The 68LC040 is a version of the 68040 that doesn't have a built-in floating-point unit, so it can't execute floating-point instructions like a 68040, a 68881, or a 68882. The emulator doesn't support floating-point instructions and it acts like a 68020 processor when processing supervisor mode instructions. In fact, this is what the Gestalt operating-system function, which Mac software can use to find out details about available hardware and software, will tell you when asked which kind of processor is installed. The emulator, however, supports a 68040-specific instruction that the 68020 doesn't: MOVE16.

The MOVE16 instruction does as its name implies: It moves 16 bytes of data in memory from the source location to the target location. MOVE16 is extremely fast on a 68040 and is used to good effect within Apple's system software. MOVE16 was implemented within the emulator as a fast memory copy, although the emulated version doesn't support MOVE16's ability to perform burst reads and burst writes to noncacheable address spaces. See Chapter 7 for more on MOVE16.

A few other details distinguish the emulator's behavior from a real silicon 68LC040. Every instruction for a microprocessor takes a certain amount of time to execute, and this time is measured in cycles. A 25MHz 68LC040 runs 25 million cycles' worth of code per second. More complex instructions generally take more cycles to complete than simpler ones. The execution times for instructions are documented by the microprocessor vendor to allow programmers to figure out the fastest way to do what they want to do, given that there are always multiple ways of accomplishing the same thing. The emulator in the Power Macs has one clear goal: to emulate as fast as possible. Because of this, instruction timings are different in the emulator than for a real 68LC040. A few 68k instructions can be emulated with a single PowerPC instruction, but most take several.

One other notable difference in the emulator is related to caching. As early adopters will remember, the introduction of the 68040 processor in Macs caused compatibility problems because of the design of the 68040's instruction cache. The Power Macs' emulator doesn't emulate the 68040 cache so faithfully that software with problems running on the 68040 will also have problems running on the emulator: in this sense, the emulator is actually more compatible with older Mac software than 68040-based Macs are.

Finally, even the emulator gets a chance to use the most fun PowerPC instruction: EIEIO. As described in Chapter 5, EIEIO (enforce in-order execution of I/O) makes sure that write operations to memory are performed in the order that the software being executed specifies. With many RISC architectures, the processor could deliberately reorder writes to memory to improve performance. In instances where writing to memory controls I/O devices, such reordering can cause big problems. The Power Macs' emulator interprets the 68k NOP instruction (no operation) and executes an EIEIO. On pipelined versions of the 68k family, such as the 68040, a NOP has the same effect as EIEIO anyway.

Floating-Point Emulation

The 68LC040 Macintosh emulator in the Power Macs explicitly doesn't emulate the floating-point coprocessor found in the 68040 chip. This has caused much consternation among some existing Macintosh users, but the omission has sound technical reasons.

First of all, the emulator *does* support floating-point math via SANE (Standard Apple Numerics Environment). Defensively written floating-point apps will first check for the presence of a floating-point coprocessor and use it if available. Then, if no floating-point hardware is available, the software should use SANE, which is guaranteed to be available on every Macintosh. Another detail not to miss is that some 68k Macs don't have a floating-point coprocessor: the LC family, the Mac IIsi, the Centris 610, and the Quadra 605, for example. So 68k software that won't run on the Power Macs for lack of floating-point hardware also won't run on these 68k-based Macs. The most precise explanation of 68k floating point on the Power Macs is that the 68881/68882/68040 processors and the floating-point instructions understood by these processors aren't emulated. This is very different from not supporting floating-point math at all.

The reason for the nonemulation of the 68k floating-point instructions can be reduced to the ratio between price and performance. The PowerPC architecture has two floating-point formats for which the floating-point hardware has explicit and optimized support: 32-bit single-precision and 64-bit double-precision. PowerPC compilers also support the 128-bit long-double floating-point format, but it's computationally more intensive, since the compilers must generate code to use the PowerPC's double-precision capabilities to mimic long double calculation and to translate between double and long double. The floating-point-capable processors in the 68k family use the 80-bit extended floating-point format. To emulate floating-point calculations using the extended format, the emulator would spend most of its time translating to and from the extended format. Since using 64-bit doubles would be less accurate than the 80-bit extendeds and unable to represent as large a range of values, the emulator would have to perform all calculations either using the long double format, or using the PowerPC's integer instructions to emulate the floating-point coprocessor. The latter is precisely what SANE on PowerPC does. Native SANE doesn't use any of the PowerPC's floating-point hardware, since it needs to provide the identical results as SANE on all other Macs, and SANE is implemented on these other Macs exclusively using integer instructions.

Emulating 68k floating-point instructions would be so computationally

> **Floating-Point Emulation (continued)**
>
> intensive that virtually no benefits would result. In addition, the engineering effort it would take to provide emulation for the full complement of 68k floating-point instructions is better spent working on other parts of the operating system to take full advantage of the PowerPC's performance.
>
> Finally, those applications that need the maximum speed provided on the 68k Macs by a floating-point coprocessor are perfect candidates to go native, since the PowerPC's floating-point performance is so high. Most floating-point-intensive applications only need single- or double-precision accuracy, so they can use one of the PowerPC's native floating-point formats to achieve significantly higher performance than 68k apps that use 68k floating-point instructions. In fact, when companies describe performance differences of two and five times for 68k versus PowerPC software, the higher number refers to the boost that floating-point-intensive applications get by running native.
>
> Floating-point math is supported in abundance on the PowerPC, even for 68k applications. Any 68k software that requires a floating-point coprocessor and that doesn't run on 68k Macs without one won't run in several existing 68k-based Macs. SANE is supported for emulated software on the Power Macs, so any software that can run on any 68k Mac will also run on the Power Macs. Finally, the high floating-point performance on the PowerPC makes going native essential for software that benefits from fast floating-point performance. Developers with floating-point-intensive software that requires 68k floating-point hardware who aren't already developing native versions of their software are showing a lack of commitment to the Macintosh.

Emulator Performance

Even with all these details about the emulator in mind, performance of emulated software still can't be predicted readily. Each piece of Macintosh software is different; Mac software spends certain amounts of time in the operating system—some software more, some less—and each piece of software calls different parts of the operating system. On the Power Macs, some parts of the OS are native and run at full PowerPC speeds, but some parts of the OS are still emulated.

This notion of a partially emulated, partially native Mac OS is called toolbox acceleration by Apple and is covered in detail in Chapters 3 and 8. Apple spent a lot of time observing which parts of the OS were most frequently used by Macintosh software. This information helped determine which parts of System 7 needed to be made native first. Obviously, those parts of the operating system where software spent the most time were prime candidates for going native. For example, QuickDraw, the Mac's graphics software, is entirely native on the Power Macs. And thanks to the wonders of the emulator and mixed mode, emulated software benefits from Native QuickDraw as well. For this reason, performance of emulated apps can't be estimated by a rule of thumb. Performance depends on how much time the software spends in which parts of the operating system and whether those parts are native or emulated. There are even parts of the operating system that exist both as 68k and PowerPC code. This is for a good reason. Since a mixed-mode switch is fairly expensive from a performance perspective, it can be faster sometimes to execute some code in emulation and avoid two mixed-mode switches. If you want to learn more about mixed-mode switches and the Mixed Mode Manager in general, see Chapter 8.

Many variables determine the performance of emulated code. The most significant factor is how much time is spent in native parts of the OS, how much is spent in emulated parts, and how much is spent in the software itself. An application such as Microsoft Excel 4.0 is a worst case for the emulator, since it spends the vast majority of time in its own code, all of which runs through the emulator, and not much time in the operating system. Such worst cases perform roughly on par with a Macintosh IIci, which is still adequate for most applications. On the other hand, 68k software that uses QuickDraw a lot will perform much better than a IIci because of Native QuickDraw's speed on the Power Macs.

How the 68LC040 Macintosh Emulator Works

The Power Macs' 68LC040 Macintosh emulator is an interpreter, not a translator. An interpreter takes every instruction to be emulated, determines what it needs to do, and then does the work. A translator takes one or more instructions to be emulated, analyzes them, and then generates native code that performs the work. The downside of the translation approach is that it generally takes more time to analyze, translate, and execute the translated code than to interpret it. However, translation has the benefit of being able to cache the native code generated by the translation process for later reuse—think of it as an emulator's Level 1 cache. Over time, the higher execution speed of often reused translated code negates the additional time taken during the initial translation.

Internally, the 68LC040 emulator uses one of the PowerPC's 32 general-purpose registers for each of the 68k's eight data and eight address registers. This direct mapping of 68k registers to PowerPC registers removes a great deal of complication for the emulator, since it doesn't have to worry about keeping track of the 68k register values.

Every time the emulator encounters a new instruction, it looks in a table that contains an entry for every 68LC040 instruction to determine what to do. Some 68k instructions, such as simple addition, can be mapped directly to PowerPC instructions. This simplifies the work that the emulator has to perform. For more complex instructions, the table entry contains a pointer to the code that will perform the work necessary to emulate the 68k instruction's behavior.

Blocks of PowerPC code that emulate a single 68k instruction can execute very fast at times. Any instruction that uses only information in registers will perform very quickly in the emulator. Any 68k instruction that must read from or write to memory will be emulated more slowly, since the PowerPC processor has to go to the trouble of looking in the Level 1 cache for the data first, then the Level 2 cache if one is installed, and finally go out to RAM to read or write the data. The large size of the 601's unified Level 1 cache is a big benefit for the emulator, since the cache stores frequently used data as well as code.

The 68LC040 emulator has some Mac-specific features built in as well. With toolbox acceleration, many parts of the Mac's operating system are already native, and emulated apps benefit from this without any action on their part. However, the emulator implements particularly performance-critical calls to the operating system directly rather than calling the OS. The amount of overhead saved and performance gained per instance is tiny, but cumulatively such optimizations can make a measurable difference. The BlockMove

> **How the 68LC040 Macintosh Emulator Works (continued)**
>
> call, an operating-system service that moves the contents of memory from one part of RAM to another, is built into the emulator.
>
> Translation is not part of the 68LC040's repertoire. Unlike both SoftWindows and IBM's Wabi, both of which perform on-the-fly translation of x86 code to PowerPC code and then store it for possible later reuse, the 68k emulator interprets each 68k instruction one by one. The emulator does, however, perform some work to look for certain common patterns in code. This allows the emulator to execute common 68k code sequences quickly.

I/O

Certain parts of the Macintosh operating system in particular still run in emulation: drivers and other software controlling input/output on the Power Macs. The Ethernet driver, sound drivers, serial drivers, and even SCSI drivers all still run in emulation. No provision exists at this writing for developing native drivers, since the 68k interrupt model differs so much from that of the PowerPC. Again, since compatibility was a primary goal, it makes sense to keep the drivers emulated. This choice also further underscores the compatibility of the emulator. If drivers run properly under emulation—not just Apple's drivers, but third-party drivers as well—this bodes well for overall compatibility.

Despite what might seem a foolish choice for performance reasons, I/O performance on the Power Macs is respectable. The main reason for this is DMA: direct memory access. As described in Chapter 2, and in more detail in Chapter 7, the Power Macs have DMA hardware that removes the CPU from the drudgework of moving I/O data through the system. The 601 concentrates on computation, and the DMA hardware handles the transport of I/O data. The drivers for the Power Macs, despite being 68k code, still take advantage of the DMA features offered by the hardware. As a result, once a particular I/O process is started, the driver generally stays out of the way until the I/O is done. On non-DMA-capable Macs, drivers are also responsible for handling writing and reading data to and from the I/O device. Such drivers would cause a big slowdown, since moving data would be handled by emulated code and not dedicated DMA hardware.

Another example where emulated drivers don't have a significant adverse effect on I/O performance is SCSI drivers. As explained in Chapter 3, the Power Macs include SCSI Manager 4.3, itself running in emulation. SCSI drivers that are 4.3-aware will reap the benefits of DMA on the Power Macs. Hard drives with 4.3 drivers perform equivalently on Power Macs and 840AVs.

The reason for this equivalent I/O performance is a straightforward one: The emulator itself is still faster than most I/O devices. In its day, the IIci was fast enough to keep up with its I/O devices, and the IIci's SCSI performance was even measurably faster than that of the IIfx, which had a 68030 running at a 60 percent faster clock rate.

One area where emulated low-level I/O software is a bottleneck on the Power Macs is networking. The protocol stacks for AppleTalk and TCP/IP both run in emulation; native versions are to be available in late 1994. The Ethernet and LocalTalk drivers that handle the work of sending and receiving raw data aren't bothered by the emulation, since the DMA hardware takes care of most of the work, but network traffic requires CPU horsepower to process information and network protocols that make up the network data. Although the performance of the protocol stacks under emulation is perfectly acceptable, even for servers, it's by no means as fast as it could be. Native AppleTalk and TCP/IP protocol stacks will speed up networking on the Power

Macs. To learn more about the new native protocol stacks and the new OpenTransport architecture, see Chapter 9.

Compatibility

The bottom line for the success of an emulator is compatibility first, performance second. A fast but flaky emulator is worse than a solid but sometimes slow one. Fortunately for Power Mac owners, the emulator is extremely reliable, and the performance of software running on the 68LC040 emulator is speedy, thanks to toolbox acceleration.

Apple's system software relies on the emulator and is integrated with it flawlessly, but the true acid test of Mac compatibility is whether the emulator can handle a power user's standard load of extensions and control panels. In tests conduced by several Macintosh trade publications, less than 1 percent of the software tested caused any compatibility problems with the Power Macs' emulator. In fact, the overall compatibility of the emulator generally exceeded expectations. Many Macintosh IS managers publicly took a wait-and-see attitude to the Power Macs, basing their skepticism on the unknown quality of the emulator.

Surprisingly enough for this industry of excessive hype, the Power Macs' emulator is rock-solid: It just plain works. Rather than fretting about possible incompatibilities, early adopters should focus their energies on acquiring native versions of the apps they use most frequently. For more information about migrating from the 68k-based Macintosh world into the universe of the Power Macintosh, see Chapters 3 and 8.

SoftWindows

Unlike the emulator built into every Power Mac's ROM, Insignia Solutions' SoftWindows is a separate native application that runs under the Mac operating system. SoftWindows emulates a complete 80286 and 80287 system as well as an MS-DOS and Windows 3.1 environment. To do this, it requires a Power Mac system with at least 16MB of RAM. All Power Macintosh configurations that come bundled with SoftWindows have 16MB preinstalled. The 80286/287 emulation and integration with DOS in SoftWindows is a direct descendant of Insignia's existing SoftPC package. SoftWindows adds the Windows 3.1 support, and it allows Windows apps to execute at low-end i486 speeds.

Insignia and Microsoft

Insignia and Microsoft entered into an agreement in 1992 that gives Insignia the license to use Windows source code directly. At the time, Microsoft was looking for 80x86 emulation technology to integrate into the Windows NT operating system. Since Windows NT runs on processor architectures other than x86, NT must offer compatibility for the existing x86 base of software until native NT apps for the different platforms become available. Currently, Windows NT is available on Alpha-based and MIPS R4x00-based systems; the PowerPC version is under development by Microsoft, Motorola, and IBM. Rather than develop its own emulator, Microsoft bought the rights to Insignia's emulation technology for Windows NT. In return, Insignia has access to the Windows source code, which allows Insignia to provide the most compatible Windows emulation possible.

Another result of access to the Windows source code is that Insignia can do the work necessary to make the most performance-critical parts of Windows run native. This is virtually identical to Apple's toolbox-acceleration strategy: The most performance-critical parts of Windows run native with SoftWindows on the Power Macs, allowing these parts of Windows to run faster than if they had to be emulated by Insignia's 286 emulator.

A further side effect of the licensing agreement with Microsoft and the resultant high degree of compatibility for applications running under SoftWindows is that Microsoft is willing to provide normal technical support for its Windows productivity applications when they are run under SoftWindows. This may seem obvious, but it isn't necessarily so. Wabi, another Windows emulation technology codeveloped by Sun and IBM, also promises to allow Windows apps to run on PowerPC systems, but Microsoft has explicitly announced that it would not support its productivity applications running on top of Wabi.

Networking

Since one of the main purposes of SoftWindows is to allow interoperability with existing applications and other x86-based machines, SoftWindows also includes networking support that allows DOS- and Windows-based applications access to network resources that they would be able to use if they were running on x86-based hardware. SoftWindows ships with full Novell NetWare support, including the IPX/SPX protocol stack and NetWare client software, so that SoftWindows users can connect to NetWare servers as if they were PCs.

No additional hardware is required for SoftWindows to act as a PC on a network. Any PC networking is routed through the Power Macs' LocalTalk or Ethernet interface or, if a separate card is installed, Token Ring is supported as well. SoftWindows includes all the necessary drivers.

SoftWindows Emulation Strategies

SoftWindows uses different approaches to maximize emulation performance. This is how it can achieve the low-end i486 performance of emulated Windows apps running on Power Macs. In contrast to Apple's 68k emulator, SoftWindows is not exclusively an interpreter. Instead, it analyzes one or more x86 instructions and translates them into PowerPC code. The analysis determines exactly what the to-be-emulated code is trying to do, and the translation phase is much like a compiler for a programming language. In this case, the translator operates on the fly, without the user having to wait perceptibly, and it generates native PowerPC code that is executed quickly on a Power Mac.

The analysis and translation can take more time than straightforward interpretation would. However, the long-term benefits of translation are great, since SoftWindows keeps the translated code in a cache, in case it's needed again. This cache functions the same way as a Level 1 instruction cache on a microprocessor, where the most recently used code is kept nearby for fast access. In the case of an emulator like SoftWindows, such a cache can boost performance a great deal, since the original time spent analyzing and translating is won back many times over if a particular piece of code is executed again frequently, obviating the need for additional analysis and translation.

The combination of cached blocks of pretranslated code and the native parts of Windows in SoftWindows makes for high performance. However, this almost-perfect picture is marred somewhat by the instruction-set architecture that SoftWindows emulates: that of the 80286. Some popular applications take advantage of features in the 80386 processor, such as its flat, nonsegmented memory model. Although Windows 3.1 itself runs fine on an 80286, the next version of Windows, 4.0, will not. It will require an 80386 at minimum.

A final part of the emulation strategy is I/O emulation. SoftWindows provides all the standard BIOS services, including support for serial ports. Any PC software wishing to use the COM1 port will never know that SoftWindows reroutes the serial traffic to and from the Mac's modem or printer port. The same goes for video: Any BIOS video calls made are converted into QuickDraw calls. Since the Power Macs all have Native QuickDraw, graphics performance for most PC apps under SoftWindows is quite high. Software that writes directly to a PC's video memory because it runs faster on PC hardware will actually run slower under SoftWindows, since the emulator has to spend time translating the code and figuring out where the software is writing to instead of just emulating the BIOS call.

i486 Emulation and Windows 4.0

Insignia Solutions has stated publicly that it intends to provide i486 emulation in SoftWindows before the end of 1994. Support for this processor in emulation is also a requirement for the next major revision of Windows, code-named Chicago.

Windows 4.0 needs at minimum an 80386 processor to run on, since it uses the 80386's flat memory model. Insignia's source-code license also extends to the source for Windows 4.0, presumably because Microsoft also wants to offer Chicago emulation for Windows NT. Consequently, Insignia is beefing up its emulator to support the i486 instruction set.

In addition to committing to providing an i486 emulator, Insignia is planning a version of SoftWindows that emulates Windows 4.0. Although it has not announced any time frames, nor has Microsoft announced any time frames for when it will release Windows 4.0, SoftWindows for Windows 4.0 will incorporate the same toolbox-acceleration features that the Windows 3.1 version has, and will also take maximum advantage of high-performance features of the Power Macs, such as Native QuickDraw.

Interestingly enough, discussion of x86 emulation on many online services often focuses on games. At this writing, ID's DOOM and LucasArts' X-WING were the most hotly debated. Both of these games require a system with at least an 80386, so SoftWindows for Windows 3.1 will not support these games. However, it's questionable whether it makes sense to run these games on an emulator in the first place. Since most of these games perform immense amounts of calculation, they spend hardly any time at all in the operating system—the worst case for an emulator. So, if you are rubbing your hands with glee at the prospect of running a 80386-or-above-only game under a future version of SoftWindows, bear in mind that this is the worst situation for the emulator, and the performance you'll get will be far less than what you would see running directly on native hardware. Then again, not many people are expected to buy an emulator such as SoftWindows to run games in the first place, and native PowerPC games should put even the highest-end x86 games to shame over time. After all, games are one area where more compute horsepower is clearly beneficial, and the PowerPC 601 and its successors have more than enough oomph for the most sophisticated games.

Wabi

Wabi started out as an acronym for Windows application binary interface. Version 1.0 of the Wabi software was designed and developed at SunSoft, the software subsidiary of Sun Microsystems, the maker of SPARC-based Sun workstations. The idea behind Wabi is similar conceptually to that of toolbox acceleration: Windows apps interact with Windows through a documented and public interface. In theory, all a Windows emulator would have to do is act like Windows and make Windows apps think they're running on top of Windows. Since this emulation would work with existing Windows apps, and not be something for developers to take into account during development, the emulator would have to emulate the application *binary* interface, the ABI, rather than the application *programming* interface, the API.

The Other Emulators

A big deal has been made about the various 68k and x86 emulation options for the Power Macs, but several lesser-known emulators run on the Power Macs as well. Independently of each other and on their own time, two engineers at Apple developed Power Macintosh–based emulators for Motorola's 6809 microprocessor. The 6809 enjoyed some success as an embedded microprocessor but certainly wasn't relevant to the personal-computer industry at any time. But many early arcade games are based on the 6809, and the point of these two emulators is to run these arcade games on a Power Macintosh. Such games should quash the notion that cool games are not available on the Macintosh once and for all.

The first of the two 6809 emulators was originally developed on a 68040-based Mac, where it ran, albeit rather slowly. Shortly after the Smurf card—the first PowerPC-based card used for early development work at Apple—was up and running, this emulator, which was written in portable C, was made to run on the Smurf. The PowerPC-native version performed much better. But the emulator alone isn't necessarily all that interesting without the games. Williams' Defender and Stargate, released in 1980 and 1981 respectively, were the first two games to run with the emulator, mainly because the hardware in the original arcade games was simple and easily emulatable. Games like Robotron and Joust use custom chips that are more difficult to emulate.

The second 6809 emulator was the result of an engineer's craving to write some sort of emulator, but not being completely sure which chip to emulate; again, the intent was to run games. Initially, a Sega Genesis emulator was considered, but abandoned. The 8-bit Z80 was ruled out, even though this CPU is used in many arcade games. The 65816, the CPU used in the Super NES and Apple IIgs, was also ruled out. The 6809 ultimately was the only one left. This second emulator was originally written in PowerPC assembly language and later recoded to portable C. The first game running on this emulator was Stargate, again because of the simplicity of hardware emulation.

The two programmers found out about each other well after both emulators were running well. However, because of the dearth of documentation about the 6809 chip, both emulators still had known bugs, although the nature of the bugs wasn't always clear. To improve both, the authors integrated the two emulators and had them run in lockstep. After an instruction was executed, each emulator would send the other the current processor state of the emulated CPU for comparison. If the states of the two

> **The Other Emulators (continued)**
>
> emulated 6809s ever differed, the developers were alerted to the problem and able to isolate it. In one instance, they even uncovered an error in early Motorola 6809 documentation. The proof of correctness of the emulator was ultimately whether they could execute the games flawlessly. This integration of the two emulators helped seek out and destroy the final bugs in both.
>
> Another emulator running native on the Power Mac—and running faster than the original hardware—is a Commodore 64 emulator. This emulator was also originally developed on a 68k-based Mac, and it was later converted to run native on the Power Macs. Although the initial incarnation of the emulator lacks support for color and sprites, it runs BASIC programs quite nicely.

A benefit of this approach is that no licensing with Microsoft is required—the interfaces from the application side to Windows are publicly documented. The downside is that Wabi can emulate only those parts of Windows that are public knowledge. This limitation hampered the first implementation of Wabi.

Another side effect of emulating just the Windows ABI is that there is no need to emulate the exact look and feel of Windows. Typical users are most familiar with the Windows user interface, but Sun's Wabi uses Motif for Windows and menus, and the Windows software is none the wiser, since the information that Wabi provides to the applications is just like Windows'.

Wabi 1.0

Sun's first version of Wabi was plagued by problems. It turned out that Windows apps, by and large, didn't exclusively stick to the documented Windows interface to do their work. It also turned out that many applications still made calls to DOS, for which there is not emulation in Wabi. Some books that explained undocumented Windows calls became popular because they explained exactly what other features were available within Windows. The downside of the proliferation of this information is that Windows developers began to take advantage of these undocumented features, making the work of the Wabi developers more difficult.

Since emulating Windows turned out to be more difficult than anticipated, SunSoft took the approach of validating and certifying Windows apps to run with Wabi. This meant that the Wabi developers did the necessary work to explicitly support individual applications and their idiosyncrasies within the Wabi emulator. A setback to the Wabi effort came when Microsoft announced that it would not support its applications running under Wabi. If a user were to call Microsoft technical support about a Microsoft application and answer "Wabi" when asked what environment the app was running under, Microsoft would tell the user that Wabi is unsupported as an environment for Microsoft applications. It's clear that Microsoft frowns upon Wabi, since it competes to some extent with Windows, but this explicit nonsupport is a setback, because Microsoft's are among the most popular and widespread Windows applications. In contrast to Wabi, Microsoft supports its Windows apps running under Insignia's SoftWindows emulator.

Wabi 2.0

IBM licensed Wabi from Sun and continued development on it. IBM Power Personal Systems, the division of IBM that will build and sell PowerPC-based desktop systems—not the popular RS/6000 family of UNIX workstations—will use a Wabi-based solution to provide the necessary emulation to help existing users of x86-based PCs migrate to their PowerPC-based systems.

However, IBM has additional emulation technologies that it will integrate with Wabi. IBM has an x86 emulator that is already shipping on IBM's RS/6000 workstations. This emulator, much like Insignia's SoftWindows, performs on-the-fly analysis, translation, and caching to achieve high performance. A distinguishing characteristic of IBM's emulator is that the version that ships with Power Personal Systems' machines will emulate the 80386 from the start. IBM will also include full emulation of DOS and BIOS in addition to the Wabi-based Windows emulation.

Since the release of Wabi 1.0, many more applications have been certified to work with Wabi. When IBM ships its PowerPC-based desktop systems in the second half of 1994, it expects to have well over 100 of the most popular Windows apps certified for use with its emulation. And since IBM has the source code to DOS and is a successful x86 system vendor, it's likely that IBM's Wabi will be a stable and solid emulation solution and will ease users' migration from x86 to PowerPC.

The Bottom Line

Emulation, in the context of the Power Macintosh, exists primarily to provide a smooth transition from 68k to PowerPC. For the companies where the Macs are the minority, consider that Windows NT will be available for PowerPC, but plain Windows never will be (this is where Insignia comes in).

The emulation solutions available on the Power Macs are reliable and perform adequately. These emulators aren't designed to compete with the top-of-the-line machines they're emulating; they provide a smooth and painless transition from the existing computing environment to the world of native PowerPC applications, which will run many times faster than their counterparts running on 68k or 80x86.

Compatibility is ultimately more important than performance, since users shouldn't have to be faced with the hassle of existing applications that don't work under emulation. In most cases, Power Mac users will be running emulated software because the native versions aren't finished yet. Upgrading 68k software to work under the emulator is frustrating and pointless, but from all appearances, it looks like this will hardly be necessary. The 68LC040 Macintosh emulator built into every Power Mac works admirably.

On the x86 side, things also look good. Insignia Solutions' x86 and DOS emulation technologies have been around for several years. The licensing agreement with Microsoft guarantees a high degree of compatibility with Windows 3.1—high enough for Microsoft to include Insignia's SoftWindows in the Windows NT operating system for RISC processors.

The Power Macs' 68LC040 emulator will provide the smooth transition that the Macintosh world needs to make the leap from 68k to PowerPC safely and without undue trauma. The performance provided by native apps will also give users of x86 machines cause to investigate, or reinvestigate, the Macintosh as a viable computing platform. With SoftWindows' ability to run popular x86 productivity software, anyone wanting to switch from the x86 world will find little reason not to switch if compelling native apps are available on the Power Macs. Even those apps available for both the Power Macintosh and high-end Pentium systems will usually run faster on the PowerPC-based machine. If such apps are available on both platforms anyway, data interchange is not likely to be a problem. All these factors come together to simplify the selection of the best tool for the task.

CHAPTER SEVEN

Power Macintosh Hardware in Depth

he raw horsepower of the PowerPC 601 chip is impressive by itself, but building a high-performance computer system involves much more than taking an existing system design and simply adding a fast CPU chip. Doing so would be like putting a 1-liter BMW boxer motorcycle engine in a Vespa scooter.

Intel experienced this phenomenon when it introduced the Pentium. The first Pentium-based systems were basically i486 motherboards with Pentiums swapped in. As a result, these first Pentium systems didn't perform much better than their i486-based predecessors, and really fast Pentium systems didn't become available until the motherboard designers designed systems to take advantage of the Pentium's features.

Despite the similarity between the Power Macintosh systems and high-end 68040-based systems, these new 601-based Macs are designed very much with the PowerPC in mind. The addition of the 601 is by no means a simple retrofit; these systems offer high performance for relatively low cost.

This chapter will provide an in-depth look at the Power Macintosh systems' hardware components as well as that of the Power Macintosh Upgrade Card, which lets users of Macs that can't be upgraded via motherboard swap upgrade to PowerPC nonetheless.

System Hardware

The Power Macs' hardware has been optimized for use with the 64-bit-wide bus that connects the 601 to the outside world. Taking advantage of this bus width provides the greatest possible throughput and permits as much data as possible to move as quickly as possible through the system.

The three Power Macs are based on a single overall design. See Figure 7.1. The Power Macintosh 6100 is the base model, since all three Power Macs share the 6100's features:

- The Power Mac 7100 is a 6100 with three NuBus slots added, and with a 601 running at a higher clock rate.
- The Power Mac 8100 has all of the 7100's features, plus a second high-speed SCSI bus, an even faster 601 than the 7100, and a preinstalled Level 2 cache SIMM.

The design of the Power Macintosh 6100, 7100, and 8100 systems can be divided into three major parts: the CPU area, DRAM, and the input/output area. These three areas are all interconnected via the Data Path chips, which are the key to much of the Power Macs' performance.

The 601 CPU Bus

Each Power Macintosh has a PowerPC 601 processor running the show. The speed of the 601 is easily discernible from the Power Mac's name: The 6100/60 has a 60MHz 601 at its heart; the 8100/80's 601 runs at 80MHz.

The Power Macs' 601 has two connections to the outside world: a 64-bit-wide data bus and a 32-bit-wide address bus. The data bus is the path over which data travels to and from the 601; the address bus is the way the 601 communicates to the rest of the system from where, or to where, data should move.

For further details about the PowerPC 601 processor, see Chapter 5.

FIGURE 7.1
The main motherboard of the Power Macs

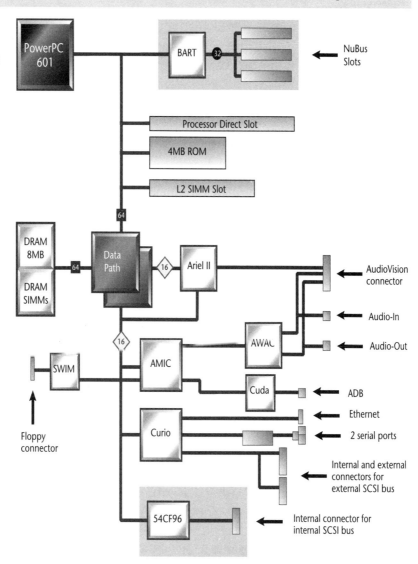

Level 2 Cache: Each Power Macintosh supports an optional Level 2 (L2) cache. The Power Mac 8100/80 comes with a 256-kilobyte L2 cache preinstalled. The purpose of an L2 cache in the Power Macs is to provide an additional buffer between the 601 CPU and the outside world: DRAM, ROM,

and the Power Macs' I/O subsystems. For a precise description of L2 caches and Level 1 caches, how they work, and why they're useful, please see Chapter 4.

With the Power Macs, the default L2 cache size is 256 kilobytes. This size was determined by a performance analysis group within Apple to offer the best ratio between price and performance. Since the high-speed static RAM used in L2 caches is relatively pricey, the measurers of performance needed to find the amount of L2 cache that would offer the highest performance benefit. As they experimented with different sizes, they found the place where the curve of performance increase flattened out at 256 kilobytes. Larger cache sizes are beneficial, but they don't offer as much of a performance boost. A 512-kilobyte L2 cache does not offer twice the performance increase of a 256-kilobyte cache. Smaller caches certainly also provide a performance boost, but the cost of the 256-kilobyte cache is small enough, considering the speed increase it offers, that it doesn't make much sense to use a size smaller than 256 kilobytes.

Unlike the Level 1 cache on the 601, the Level 2 cache in the Power Macs has only one mode: write-through. This means that any data written to the L2 cache is written out to DRAM, or I/O space, as quickly as possible. In contrast, parts of the 601's L1 cache can be defined as copyback, where the data *isn't* written back to DRAM, or the L2 cache, immediately. Copyback mode is somewhat faster than writethrough because it causes less bus traffic.

High-Speed Memory Controller: The high-speed memory controller (HMC) in the Power Macs is the nexus for interactions between the 601 CPU and memory. The HMC is designed to provide for the 601's memory-access needs, as well as connections to ROM, L2 cache, DRAM, the processor direct slot (PDS), and the Power Macs' input/output (I/O) devices.

Although the 601 and the DMA logic in the Power Macs actually write to or read from memory, HMC keeps track of memory being read from or written to and makes sure that no two parts of the system try to get at the same part of memory at the same time. It is, in a sense, a traffic cop for the data moving throughout the Power Mac system.

This function is particularly important because the design of the I/O on the Power Macs is memory-mapped. This means that writing data to a particular I/O device is just like writing data to a particular address in RAM. The HMC, in combination with another chip called AMIC, makes sure that all the right things happen and that data goes where it's supposed to.

HMC supports a maximum of 264MB of installed RAM. To install this much RAM in a Power Mac would require a 32MB SIMM installed in each of the Power Macintosh 8100's eight SIMM slots. On the Level 2 cache front, HMC supports sizes starting at 128 kilobytes; the upper limit is determined only by hardware technology. When the Power Macs' design was being developed, it was assumed that 512 kilobytes of L2 cache would be the largest possible. 1MB L2 caches for the Power Macs are already advertised today.

When HMC plays traffic cop on the buses of the Power Macintosh, it has to decide who gets access when multiple parties want use of the bus at the same time. DRAM refresh gets the highest priority; for dynamic RAM to keep its contents intact, it has to receive a signal every so often. If the refresh doesn't happen, data is lost from memory. Video gets the next highest priority on the bus, which is important because video information destined for a monitor is time-critical. Since the motherboard-based video subsystem uses system DRAM to store video data, it's possible that the CPU may stall when wanting to read some part of DRAM while the video data is accessed. If this happens, the CPU must wait until the video subsystem is done fetching the video data from DRAM before it can perform its DRAM access. As a result, the CPU gets last priority, since it is assumed that it can wait a bit, whereas the other two types of access to DRAM are time-critical and have to happen *then* rather than later.

The order of priority in the Power Macintosh 6100, 7100, and 8100 is as follows, from the highest to lowest:

Order	Device
1	DRAM Refresh
2	Video Refresh
3	I/O DMA
a	Sound
b	SWIM III
c	SCC
d	Ethernet
e	SCSI
4	Processor Direct Slot
5	The PowerPC CPU

ROM: Each Power Macintosh has 4MB of ROM. ROM has traditionally been slow to read from—much slower than DRAM. The ROM in the Power Macs consists of 120ns burst-mode ROM—slower than DRAM, but not so slow as to cause a significant performance impact on the system. Faster ROMs would offer an insignificant performance improvement but increase the cost a great deal.

The path between the Power Macs' ROM is 64 bits wide. This path allows the Power Macs to access ROM data as quickly as possible, which is important because much of the system software is contained in the Power Macs' ROM and needs to be accessed frequently.

Processor Direct Slot: The processor direct slot (PDS) on the Power Macs is a direct connection to the 64-bit CPU bus. Any card installed in this slot behaves no differently than if it were hard-wired into the system.

Although Apple has documented the PDS in the Power Macs, it's not encouraging developers to design cards for it, for the following reasons:

- Too many different kinds of slots are available on the Macintosh as it is; the last thing Macintosh card developers need is another type of slot to support.
- There really shouldn't be a need for third-party developers to build PDS cards for the Power Macs. Most existing NuBus cards work fine, and most standard expansion-card functionality, such as Ethernet, is already supplied on the motherboard.
- If many third parties jump on the Power Mac PDS bandwagon, Apple will have to provide long-term support for this PDS design. The current PDS interface is 601-specific and would require some changes for 603- and 604-based systems, so it wouldn't be a general-purpose solution.
- Apple has already announced its intent to support the PCI (Peripheral Component Interconnect) expansion-card standard in the second generation of PowerPC-based Macs. PCI has significantly more bandwidth available than NuBus—at least three times as much. Since it's already known that a better, faster, multiplatform standard expansion-card interface is on its way, there's no good reason to create a new one.
- The only very high bandwidth application for the Power Macs that requires a bus faster than NuBus is audio and video, and the AV Card for the Power Macs supports the DAV digital audio/video slot originally introduced with the Quadra 660av and Quadra 840av.

- The Power Macintosh 7100 and 8100 models come with either the VRAM or the AV Card preinstalled in the PDS. The non-AV version of the 6100 is the only Power Mac that ships with an unused PDS, which is sometimes filled with the NuBus adapter.

Given all these reasons, it doesn't make sense for the PDS to be public. As a result, buyers of expansion cards for the first generation of Power Macs will be limited to NuBus cards. The only thing to consider when buying NuBus cards for the Power Macs is that the cards should come with software drivers that specifically support the Power Macs. Older drivers will work, but not necessarily as quickly. Some existing drivers will cause the Power Macs to slow down, since the drivers are designed with 68k Macs in mind; depending on how they interact with the rest of the Macintosh system, they can cause many mixed-mode switches. Chapter 8 explains the issues surrounding this problem.

BART: BART is the name of the Power Macs' NuBus controller. The Quadra 660AV and Quadra 840AV contain a NuBus controller named MUNI, short for Macintosh Universal NuBus Interface. (MUNI is also the name of San Francisco's light-rail system.) When the Power Macs' NuBus controller chip needed a name, the result was also the namesake of a local public transportation system: BART (short for Bay Area Rapid Transit, which serves much of the San Francisco Bay Area).

BART provides the same functionality as MUNI does in the AV Quadras, but with a PowerPC twist. BART's interface to the rest of the Power Mac system is a full 64 bits wide; it can support one and four beat transactions between NuBus and the Power Macs' CPU bus, which move 64 or 256 bits at a time, respectively.

NuBus Performance

Even though the Power Macs have the most popular features found on NuBus cards built in, there are still many good reasons to use NuBus cards. However, two major issues can affect NuBus performance on the Power Macs. The most popular NuBus cards used on the Power Macs tend to be high-performance cards such as video frame grabbers or QuickDraw accelerators. These cards need every little bit of NuBus performance so they can transfer data and perform their work as quickly as possible.

Excessive Mixed-Mode Switches

NuBus cards need driver software for the Mac to be able to use them. In the case of some NuBus cards—most notably, QuickDraw accelerators—the driver software redirects parts of the OS to use the driver to perform certain operations. Existing NuBus drivers are written in 68k code, which causes no problems until this code is called frequently by PowerPC code. QuickDraw is native on the Power Macs, so any QuickDraw accelerator drivers that intercept QuickDraw calls will cause a major slowdown because they're 68k code. To remedy this, vendors of NuBus cards whose drivers need to intercept native PowerPC code must supply new versions of their drivers to eliminate the frequent switches between PowerPC code and emulated 68k code.

Bursting into the Bus

The other issue with NuBus on Power Macintosh is hardware-related, but is partially solvable by software. Since the Power Macs are supposed to be the fastest Macs available, they are commonly compared with the top-of-the-line 68040-based Mac, the Quadra 840AV, which has the highest NuBus performance of any Macintosh. On the 840AV, Nubus drivers can take advantage of a particularly useful 68040 instruction: MOVE16. This instruction moves 16 bytes of data via burst reads and writes, achieving the fastest transfer rates to and from NuBus cards on any Macintosh.

The PowerPC 601 processor has no similar instruction to MOVE16. Although it is certainly possible to move 16 bytes during one burst to and from the 601, it's not possible to do this between the 601 and NuBus. The 601 is unable to burst-read or burst-write to an address that is not designated as cacheable memory; the 68040 can.

As you will remember from Chapter 4, several modes of caching exist. For addresses that are I/O devices, it is a bad idea to mark this memory as cacheable, since a memory-mapped I/O device won't behave like a memory address. If you write a value to an I/O device's address and read back from that address,

> **NuBus Performance (continued)**
>
> you often won't read the same value that you just wrote, especially since I/O addresses often deliberately behave differently depending on whether they're being read from or written to.
>
> Because of this disparity between the written data and the value resulting from immediately reading from that same address again, marking such an address as cacheable is dangerous, since an immediate read after writing to a cacheable address will return a value from the cache rather than the real data from the I/O device.
>
> Since the 601 can't burst to or from noncacheable addresses, and the 68040's MOVE16 does allow bursts to and from noncacheable addresses, the 601's superior overall performance won't make up for the lack of this particular feature of the 68040.
>
> To help remedy the situation, Apple has provided developers with a new call, PBBlockMove, which they can use to move memory to and from I/O addresses as quickly as possible. Since this API is independent of how the memory copying is done, Apple's engineers can continuously improve PBBlockMove's performance over time without driver writers ever having to know about the details. The initial version of PBBlockMove provides performance that exceeds that of any Mac except the 840AV. Future versions of PBBlockMove may very well reach or eclipse the 840AV's NuBus transfer rate.

The NuBus 90 slots in the 7100, 8100, and the 6100's NuBus adapter card are functionally identical to those found in the Quadra 840AV and on the Quadra 660AV NuBus adapter card. The NuBus 90 specification supports standard 10MHz operation for a maximum throughput of approximately 38MB per second. NuBus 90 also has a 20MHz burst mode for high-speed transfers, but only between cards. The maximum throughput of this mode is roughly 76MB per second. These speeds are the same as in previous Macs. Some applications, notably audio and video, have often been thwarted by NuBus' throughput. This doesn't change for the Power Macs and it won't really be addressed until a PCI card standard appears in future Power Macs. If you'd like to know more about PCI, Chapter 9 covers it in greater detail.

Data Path

The two Data Path chips in the Power Macs hold the key to the overall high performance exhibited by these new PowerPC-based machines. Data Path is divided into two chips for a relatively banal reason: production cost. One of the chips handles the even bits on the bus, the other handles the odd ones, and it's simply cheaper to produce two chips rather than a single huge chip.

Data Path has four different connections to the different buses:

- DRAM bus: 64 bits wide
- CPU bus: 64 bits wide
- The pixel bus leads to Ariel, the chip responsible for the motherboard video subsystem: 16 bits wide
- I/O bus: 16 bits wide

The Data Path chips create four almost-autonomous subsystems within the Power Mac hardware, and each subsystem can do its thing without adversely influencing the others. The Data Path combined with DMA allows high-speed I/O while high-speed processing continues unaffected.

Data Path is crucial to the Power Macs' performance because it acts as a router, directing data so as not to bother any part of the system unnecessarily. The Data Path chips separate the CPU bus, the I/O bus, DRAM, and the video subsystem from each other. With this division in place, data coming from, for example, a serial port can be moved to DRAM without the CPU bus being affected, allowing the 601 to continue its work uninterrupted and allowing the use of the full bandwidth of the CPU bus all the while. The 601 could be writing graphics data to a video card while the serial transfer goes on, and neither is affected by the other.

As part of its duties to separate the CPU bus from the rest of the system, Data Path also contains write buffers so that any write transactions from the CPU bus finish as quickly as possible. The idea is similar to the idea behind Print Monitor, which attempts to give you back control of your Mac as soon as possible, and then performs the printing while you're doing other things. For example, when the 601 writes to memory, the Data Path will accept the data and make sure that it makes it to DRAM. In the meantime, the 601 can get on with its work. This is a further contributor to the high system performance of the Power Macs.

SCSI and Ethernet I/O are also noteworthy since Data Path contains special buffers for these two ports. Since speed is of the utmost importance with these two, Data Path performs a process called *byte assembly* on data coming from either of them. Rather than sending each byte of data as it arrives from the SCSI or Ethernet port, Data Path waits for 8 bytes, or 64 bits, to collect before it moves the data across the bus to DRAM in one go. This is far more efficient than sending eight individual bytes across the bus; moving one 64-bit chunk requires only a single bus transaction rather than eight, even though the amount of data is the same. Byte assembly happens for data received from these two ports because they have the highest throughput. Sound data is byte-assembled as well. This scheme also works for these three ports in particular: Since they're used frequently, it's reasonable to assume that once a byte arrives from one of these ports, more data will follow shortly. So it's worthwhile to wait a bit to get the whole 8 bytes together. For slower I/O ports—for example, ADB—byte assembly makes no sense at all, since ADB data comes along far less frequently.

DRAM

Each Power Macintosh has 8MB of 80ns DRAM soldered on the motherboard. Each of the three models has different numbers of SIMM slots for RAM expansion: two in the 6100, four in the 7100, and eight in the 8100.

A single 64-bit path goes into DRAM from the Data Path. This is the only way in or out of DRAM. The width of this bus is important, since it determines the maximum amount of data that can go in and out of RAM during a single transaction. Since the Power Macs use the standard 72-pin SIMMs also used in the Quadra 650, Quadra 800, Quadra 660AV, and Quadra 840AV, which only support 32-bit-wide access, two identical SIMMs must be added at a time to expand a Power Mac's RAM.

When you are using the on-board video, 600 kilobytes of the Power Macs' DRAM are set aside for the video buffer. On top of the memory taken up, and thus lost to application software, use of system DRAM for video also consumes bandwidth on the DRAM bus. Monitors need to receive updates to their information many times per second; typical monitor refresh rates are 60 or 75MHz. This means that 60 to 75 times per second, up to 600 kilobytes of data is moved from DRAM and sent out to the monitor via the on-board video subsystem. This is approximately 37MB per second of video data, and it uses up a significant fraction of the bus bandwidth. Here again, Data Path's isolation of the Power Macs' different buses allows individual parts of the system to run as efficiently as possible.

The I/O Bus

The Power Macs' input/output devices are connected to a 16-bit bus that goes to the Data Path. The one exception is NuBus: The BART controller is on the CPU bus, since transfers to and from NuBus require significantly more bandwidth than even the highest-speed SCSI transfer. Since NuBus is 32 bits wide, it's much closer to the CPU bus in performance than to any of the I/O devices.

Although a 16-bit-wide I/O bus might appear to be small, it is sufficient for the type of transfers it needs to support. Even on the Power Macintosh 6100/60's 30MHz bus, the slowest of the three Power Macs' buses, the 16-bit I/O bus supports up to 57MB per second. The 8100/80's I/O bus supports approximately 76MB per second, more than enough for the standard 5MB per second SCSI bus and the 8100's Fast SCSI high-speed internal bus, with plenty left over for other I/O. For example, the theoretical maximum bandwidth required for Ethernet is 1.25MB per second, but it typically uses only around 500 kilobytes per second on most computer systems. Each GeoPort serial port is capable of 256 kilobytes per second throughput, and CD-quality audio only takes up about 86 kilobytes per second. As these numbers show, the 16-bit I/O bus has ample available bandwidth.

The I/O bus and the different I/O devices all provide comparable performance to the fastest 68040-based Macs. The on-board video, despite using system DRAM and precious bus cycles, outperforms many other video solutions on the Mac market. The VRAM Expansion Card, combined with Native QuickDraw, offers performance that rivals some QuickDraw accelerator cards, and the Power Macintosh AV Card adds video capture and playback features to these powerful systems.

AMIC: AMIC is short for Apple memory-mapped I/O controller. AMIC controls the data flow between the different I/O devices and the rest of the system and manages the direct memory access for SCSI, Ethernet, audio, the two serial ports, and the floppy drive.

AMIC also handles interrupts generated by the different I/O devices. This is a critical feature for the Power Macs, since throughout the history of Macintosh hardware architecture, the 68k family's interrupt scheme has been at the core of hardware I/O.

Bandwidth

A great deal of data flows through a Power Macintosh system. The purpose of the Data Path is to isolate the four different buses inside the Power Macs (CPU, DRAM, Video, and I/O), so that data flow affects only the buses that it has to, thus making the most bandwidth available on every bus.

Table 7.1 gives the available bandwidth of the various buses on a Power Macintosh 6100/60 as well as the maximum bandwidth requirements of various I/O devices. The buses of the 7100/66 run at 33MHz; the 8100/80's run at 40MHz.

The 20MHz NuBus block transfers go only from card to card, never between a card and the rest of the system. As a result, a NuBus transfer to the Power Mac system could never use more than 38.15MB, which is the theoretical maximum of NuBus. In reality, the implementation of NuBus, including the overhead of NuBus protocols and the time taken to synchronize NuBus to the system bus, results in real NuBus throughput of between 7 and 15MB per second, depending on whether data is going to or coming from NuBus, and which Power Mac is being used.

Table 7.1 Bandwidth

Available Bandwidth (6100/60)		
CPU and DRAM Bus	30MHz x 64 bits	228.88MB/sec
I/O and Video Bus	30MHz x 16 bits	57.22MB/sec
Bandwidth Used		
Video Data	75Hz x 600kB	43.95MB/sec
Internal SCSI (8100 only)	hardware max	10MB/sec
External SCSI	hardware max	5MB/sec
Ethernet	10 Mbps (theoretical max)	1.25MB/sec
GeoPort	2 Mbps (max)	0.25MB/sec
CD-Quality Audio	44.1kHz x 16 bits	86.25kB/sec
LocalTalk	230.4 kbps	28.8kB/sec

Interrupts are signals generated by I/O devices that cause the main processor to stop what it's doing and deal with the cause of the interrupt. I/O is time-critical and can't be handled whenever the CPU gets around to it. The 68k family has seven levels of interrupts, which signify seven levels of importance. Different I/O has a different interrupt level, and thus a different interrupt priority. A higher-level interrupt can interrupt the processing of a lower-level interrupt but not vice versa.

The 68k offers a comparatively rich set of interrupt levels, but the PowerPC only has a single one. So it's up to AMIC to handle the simulation of the different interrupt levels for the different I/O devices. This simulation is necessary to make the I/O subsystem as compatible with previous Macs as possible. Had the designers of the Power Macs not made this effort, no existing Macintosh peripherals would work with the Power Macs unless all their drivers were completely rewritten.

Curio: The Curio chip is the most versatile of the I/O chips in the Power Macs. Curio is responsible for handling SCSI, Ethernet, and both serial ports for the Power Macs. Curio is also used to support the same ports in the 660AV and 840AV.

The SCSI port that Curio is responsible for can support throughput of up to 5MB per second.

The Ethernet part of Curio connects to the outside world through the Power Macs' AAUI interface, which is a medium-agnostic Ethernet port. To actually connect a Power Mac to an Ethernet network, you need an adapter for either 10BASE-T (twisted-pair), 10BASE-2 (coax/ThinNet), or 10BASE-5 (AUI/ThickNet) Ethernet wiring.

Finally, Curio contains the hardware for the two serial ports in the Power Macs. Each port has 8 bytes of buffer when LocalTalk and GeoPort are active; otherwise the serial buffers are 3 bytes for incoming and 1 byte for outgoing

GeoPort

GeoPort is a new type of high-speed serial port that was originally introduced with the Quadra 660AV and Quadra 840AV. It started as an Apple-only innovation, but in early March 1994 was licensed to Aox and Analog Devices. These two companies will make the GeoPort technology available for the x86 hardware as well as the Windows and OS/2 software environments.

A GeoPort system has multiple parts: the GeoPort hardware built into a computer system, an external GeoPort adapter, and the necessary system software to allow the computer to access and control the adapter. The GeoPort port itself looks like a standard mini-DIN-8 serial port, just like the modem and printer ports on Macs since the Mac Plus. The only difference is that a GeoPort has an additional ninth pin, which is used to supply power to an external GeoPort adapter. The GeoPort port is designed so that all existing DIN-8 plugs will work without modification.

The GeoPort can act as a conventional modem or printer port, allowing up to 57600bps asynchronous serial rates as well as supporting LocalTalk's 230.4kbps data rate. When communicating with an external GeoPort adapter, the GeoPort port operates at 2 megabits per second, enough to support very high bandwidth serial applications.

The first GeoPort adapter available was Apple's own Telecom Adapter, which offers two RJ-11 jacks to plug standard phone wires into. The Telecom Adapter contains the hardware necessary to convert the analog signals from a phone line into digital data, and vice versa. The adapter has no modem capabilities; it provides the minimal hardware necessary to connect a system to phone lines.

The GeoPort for Power Macintosh software provides all the features and functions of a modem purely in software. Combined with the Telecom Adapter, the GeoPort software acts just like a stand-alone modem, but at a considerably lower price. Version 1.0 of the GeoPort for Power Macintosh software provides the functions of a 14.4 kilobits per second V.32bis modem for data as well as that of a 9600bps V.29 fax modem. In addition, the GeoPort software comes with Apple's Express Modem software, which makes the GeoPort look like a regular modem to conventional communications software.

The software modem runs on the Power Macs' 601 chip, which illustrates clearly how much computational horsepower the Power Macs have. Both the 660AV and 840AV included a separate DSP (digital signal processor) chip that, combined with the appropriate GeoPort software, ran a software modem on these

> **GeoPort (Continued)**
>
> machines. The Power Macs offer the same features, but without the need for a separate DSP chip.
>
> Another benefit of a software modem is that it's upgradable over time. When the GeoPort modem was first introduced for the 660AV and 840AV, it was only able to support 9600bps for data connections. Shortly thereafter, a software upgrade made 14400bps available to users, without any change to the Mac or the GeoPort Telecom Adapter.
>
> Current GeoPort is usable only with Apple's own Telecom Adapter, but future plans call for an ISDN GeoPort adapter, as well as adapters for T-1 high-bandwidth digital phone lines, which provide up to 1.5 megabits per second throughput. To use one of these alternative connection methods, all that's needed is a new adapter, since most of the necessary hardware is built into each Power Mac.

traffic, as on most other Macs. The combination of the larger buffer and the systems' support for DMA for serial I/O make the two serial ports capable of supporting Apple's GeoPort high-speed serial architecture.

AWACs: The AWACs (audio waveform amplifier and converter) chip is in charge of all audio-related I/O in the Power Macs. AWACs is a further evolution of the Singer audio chip and waveform amplifier chip (WAC), both of which are found in the Quadra 660AV and Quadra 840AV.

AWACs has three stereo inputs and internally supports two channels of 16-bit sampled digital audio. The protocol used by AMIC and AWACs supports eight channels of 20-bit data.

SWIM III: The SWIM III chip is the Power Macs' floppy controller. SWIM is a spoonerized acronym that stands for super-integrated Woz machine; the original integrated Woz machine was a floppy controller designed by Steve Wozniak, cofounder of Apple Computer. The original SWIM included support for the 1.44MB high-density floppy format. SWIM III is the third generation of the SWIM chip.

SWIM III supports DMA data transfer of floppy data and doesn't need interrupts to be disabled during floppy I/O. In the past, use of floppies has been unfriendly in the Mac hardware architecture. During floppy access, virtually all other hardware I/O stopped until floppy I/O had completed, since the CPU was responsible for moving the data to and from the floppy. With the added support for DMA, the CPU must no longer babysit the floppy controller during I/O and is free to do real work.

Ariel II: The Ariel II chip combines a color lookup table (CLUT) and a digital-to-analog converter (DAC) in a single chip. Ariel II is the same chip used for on-board video in the Macintosh Color Classic. Ariel II contains a 256-element color lookup table and the necessary circuitry to convert the video data in DRAM to analog signals for the on-board AudioVision port.

Ariel has two connections to the rest of the system. One, the pixel bus, is to the Data Path chips, where video data comes from DRAM. The other connection goes to the Power Macs' I/O bus and is used to set up and control Ariel's functions.

CUDA: The CUDA chip is responsible for managing the Apple Desktop Bus (ADB), turning system power on and off, managing Parameter RAM, and managing the built-in clock.

Input devices such as keyboards and mice use the ADB to communicate with the Macintosh system. Since the 7100 and 8100 systems are turned on via the keyboard and turned off by software, combining ADB and the on/off switch in the same chip makes sense. In addition, GeoPort devices can also turn on a Macintosh. This allows incoming phone calls, or an incoming fax, to start up the Power Mac and allow it to receive incoming data.

Squidlet: The Squidlet chip provides all the clock signals needed by the CPU and other ASICs within the Power Macs except for the clock signals for on-board video.

Squidlet provides 2x and 4x clocks for the 601 CPU, as well as 2x and 1x clocks for the other ASICs. Squidlet's main clock speed is determined by the Power Mac system it's in; on the 6100/60, for example, Squidlet's main clock runs at 30MHz, hence the 60 and 120MHz clocks on the 601.

54CF96: The 54CF96 is the second SCSI controller in the Power Macintosh 8100. Curio handles the standard external SCSI bus in the 8100, and the 54CF96 is in charge of managing the internal high-speed SCSI bus that can support up to 10MB per second SCSI throughput. The 53CF96's bus is accessible only via a ribbon connector on the 8100's motherboard.

Upgrade Card

The Power Macintosh Upgrade Card is the lowest-cost upgrade solution for owners of 68040-based Macs. Only some Quadra models can be upgraded to one of the Power Macs via logic board upgrade. Some Macs can't be upgraded in this way, but Apple didn't want to leave these machines in the lurch.

The Power Macintosh Upgrade Card contains a PowerPC 601 chip, 1MB of 15ns Level 2 cache, and a standard 4MB Power Macintosh ROM. The 601 always runs at twice the speed of the system it's installed in. For example, in the 25MHz 68040-based Quadra 700, the 601 on the Upgrade Card runs at 50MHz.

The 601 used on the Power Macintosh Upgrade Card is slightly different from the ones found inside the Power

Macs. The difference between the 601s is a minor one; Upgrade Card users will almost certainly never notice.

The 601s on the Upgrade Card run their floating-point units in so-called synchronous mode. This mode defeats some of the benefits offered by pipelining in the floating-point unit. For this reason, floating-point performance on Upgrade Card–based Power Macs will be lower than on equivalently fast 601s in Power Macs. This decision was made deliberately, since those users who need the utmost in floating-point performance will almost certainly buy a new Power Mac anyway. Users upgrading via the card will still see floating-point performance far higher than on their 68040 Macs, but not quite as high as on the Power Macs.

The Upgrade Card runs at twice the speed of the system it's installed in because it makes the work of the bus converter on the Upgrade Card much easier. For the Upgrade Card to function properly, it must contain a converter that allows the 601 bus on the Upgrade Card to work with the 68040 bus on the Macintosh system the Upgrade Card's plugged into. Since the Upgrade Card contains no RAM of its own, nor any I/O ports, it's dependent on the host system for these resources. The large Level 2 cache acts as a buffer to counteract much of the performance hit caused by having to access the slower 68040 bus to get at RAM and I/O.

AV Card

The Power Macintosh AV Card is the only part of the Power Macintosh hardware that is basically identical to existing hardware on a 68k-based Mac. The AV Card contains virtually the same video hardware found in the Quadra 660AV and 840AV. The audio features in the AV-equipped Quadras are available on all Power Macs, not just those with the AV Card.

The Power Macintosh AV Card uses several different ASICs to do its job. Since the design of the card is an adaptation of the design in the AV Quadras, the AV Card contains a bus-converter chip named PODRIC that provides the necessary signal conversion and buffers to translate the Power Macs' 601 bus signals into 68040 bus signals that are understood by the chips on the AV card.

All of the AV Card's sound capabilities come from the AWACs chip on the Power Macs' motherboard; the AV Card itself has no additional audio hardware.

CIVIC

The heart of the AV Card is CIVIC, the Cyclone integrated video controller. Cyclone was the code name for the Quadra 840AV, and CIVIC was originally designed for the two AV Quadras. CIVIC can manage between 1 and 4MB of VRAM, although the AV Card has 2MB soldered on it and has no expansion capabilities. CIVIC also controls the interaction between the Philips SAA7194 chip and the Sebastian chip, and it provides timing signals for the different standard television formats. It additionally is responsible for handling the convolution of graphics for line-interlaced displays such as televisions, which allows the AV Card to be used to display Macintosh video data on TV screens.

Sebastian

Sebastian is a combination CLUT and DAC, similar to Ariel II, but designed for higher performance. Sebastian has two 32-bit-wide connections to the rest of the card. It can accept data either as a 64-bit quantity coming in both ports or as one or two individual 32-bit parts.

Sebastian allows one of its 32-bit ports to be used for digital video while the other is processing graphics data such as QuickTime. This feature makes it possible to mix video and graphics on the same screen, even if the two have different bit depths.

SAA7194

The SAA7194 chip is a single-chip version of the two-chip set used in the Quadra 660AV and Quadra 840AV. The SAA7194 is made by Philips and is used on the AV Card to decode video data from the incoming video port in either S-video or composite NTSC, PAL, or SECAM format and translate the analog data to a digital format.

The SAA7194 chip also provides the ability to scale the incoming video picture in hardware. No computationally intensive and therefore slow software scaling is required. When the Philips chip is done decoding the incoming video signal, it passes the digitized data into VRAM as either 16-bit RGB data, 8-bit grayscale, or YUV.

Mickey

Mickey is the chip responsible for video output. Outgoing video can leave the AV Card either via a standard DB-15 monitor connector or via the outgoing S-video connector.

Mickey can output video data as RGB, or it can translate the RGB data into composite NTSC, PAL, or SECAM format as well as S-video.

The DAV Connector

A new type of connector was introduced with the 660AV and 840AV to allow NuBus card developers direct access to the raw audio and video data in the two AV systems. This connector was named DAV, for digital audio/video.

A typical use for the DAV connector is hardware-assisted video compression and decompression, to allow larger-size video windows and higher frame rates than software-based compression schemes. The DAV slot allows the NuBus cards direct access to the digital data without having to transfer the data via NuBus.

In the AV Quadras, the DAV slot is inline with a NuBus slot in the system. In the case of the 660AV, the DAV slot is on the NuBus adapter card that plugs into the 660AV's PDS slot. Either way, the only realistic way to connect to the DAV slot is from a NuBus card that has its DAV and NuBus connectors lined up to plug into both connectors in the Quadras. Since the video part of the AV features for the Power Macs is implemented on a card, it doesn't make sense to put a DAV slot on the motherboard, inline with a NuBus slot on the 7100 or 8100. The 6100AV models are particularly tricky in this regard: A 6100AV cannot support a NuBus adapter card, since the AV Card is plugged into the only processor direct slot, the same slot that would otherwise house the 6100's NuBus adapter card.

The DAV connector for the Power Macs is designed to be connected to a DAV connector on a NuBus card via a ribbon cable. Since the 6100AV model can't support a NuBus card in addition to the AV Card, the DAV slot is something of an atavism in this system. The electrical signals for the ribbon-cable version of DAV are identical and provide the same access to the raw digital audio and video data as the DAV slots in the 660AV and 840AV.

VRAM Expansion Card

Those models of the Power Macintosh 7100 and 8100 that don't have an AV Card come with the VRAM Expansion Card preinstalled in the processor direct slot. The VRAM Expansion Card is a VRAM-based frame buffer that provides significantly higher video performance than the video subsystem on the Power Macs' motherboard. The VRAM Expansion Card consists of the VRAM, a video/VRAM controller, a DAC (digital-to-analog converter), and a clock generator.

Two versions of the VRAM Expansion Card are available: The card that comes with the 7100 has 1MB of 80ns VRAM soldered onto the card itself, with four SIMM slots to allow expansion up to 2MB of VRAM. The 8100 version of the VRAM Expansion Card comes with 2MB soldered, with SIMM slots for an additional 2MB of VRAM. Other than the preinstalled VRAM and its expandability, the two cards are completely identical.

The VRAM Expansion Card contains

- DaMFB, the dual-array memory frame buffer chip
- RaDACal, a color lookup table and digital-to-analog converter (CLUT/DAC) designed specifically for this card
- PUMA, a clock-generator chip

DaMFB acts as the memory controller for the VRAM on the card, managing VRAM refreshes and access to the VRAM data. RaDACal is a combination CLUT and DAC, much like Ariel for motherboard video and Sebastian on the AV Card, except that it supports up to 24 bits per pixel and is designed specifically for the 64-bit-wide bus on the VRAM Expansion Card. PUMA is the chip that generates all the timing signals on the VRAM Expansion Card, for VRAM refreshes as well as for video refreshes for the monitor.

The VRAM Expansion Card has a single, standard DB-15 monitor interface, the same DB-15 connector that Macintosh users (and their monitors) have been accustomed to since the introduction of the Apple 13-inch RGB monitor.

CHAPTER EIGHT

Power Macintosh Software in Depth

he decision about whether to make code native is not a simple matter of saying "yes!" and then doing it. It requires careful consideration about the type of code involved and the work it performs. In general, most applications benefit from going native. Those applications that spend most of their time waiting for I/O to happen benefit less than others that spend the majority of their time performing computations. In all cases, however, user interfaces can benefit greatly from the added performance, so developers of I/O-bound apps should also try to go native. One of the Mac's main benefits is the high quality of its user interface. With the introduction of the PowerPC, much more interactive and responsive user interfaces are possible.

Mixed Mode

The Mixed Mode Manager handles all the work involved in switching between native and 68k code on Power Macs. Since existing 68k software has to run unmodified on the Power Macs, the Mixed Mode Manager is completely transparent to 68k code. This means that so-called accelerated toolbox traps—parts of the operating system that are native—appear no different to 68k software than emulated ones. However, this means that the Mixed Mode Manager is handling the switch from 68k code to PowerPC behind the scenes and also handling the switch back when the native code is done and control needs to be returned to the 68k software.

Mixed-Mode Switch: 68k to PowerPC

The exact mechanics of a mixed-mode switch from 68k to PowerPC code varies somewhat, depending on whether the 68k passes any additional information to the native code in the form of parameters. When one 68k routine calls another, it must adhere to standard calling conventions that define how the caller and the called routine exchange information. Both parties in the exchange need to know exactly how much information is passed from one to the other and how it is passed. Software written in Pascal passes parameters differently than software written in C. Most Mac OS routines use the Pascal calling conventions; there are exceptions to this rule that have calling conventions of their own, which conform to neither those of C nor those of Pascal.

When a piece of 68k code calls native code, a piece of information known as a *routine descriptor* is used to make the transition. The routine descriptor, a data structure introduced with the new runtime architecture of the Power Mac, contains all the information that the Mixed Mode Manager needs to know about the routine that's being called. The Mixed Mode Manager must convert and possibly reorder any parameters that the emulated software passes to the native routine, before the native software is executed. In addition to the information about the called routine, the routine descriptor contains a pointer to the *transition vector*. This data structure contains two pointers:

- A pointer to the actual PowerPC routine
- A pointer to the called routine's global variables

Mixed-Mode Switch: PowerPC to 68k

When the native code is done executing, any return values need to be converted by the Mixed Mode Manager into the form that the emulated 68k code expects. The Mixed Mode Manager adheres to the standard register-saving conventions on the 68k side:

- Registers whose contents used to be documented as saved will be saved
- Registers whose contents could change almost certainly will change

One of the sneakiest pitfalls of the switch from 68k to PowerPC-based machines happens if the 68k software relies on undocumented behavior of system calls. If, for example, a toolbox call on a 68k machine deposited an undocumented but useful value in a specific register that is not guaranteed to be saved, you should expect that this value may no longer appear there. Any software that depends on such undocumented behavior will malfunction on the Power Macs—if not now, then maybe after the next system software release or with the introduction of new PowerPC-based machines.

Penalty for Switching

No matter in which direction a mixed-mode switch occurs, the switch takes time. The average mixed-mode switch takes about 50 emulated 68k instructions. Although this may seem short in human terms, this is a significant lag for a computer. The cumulative effect of many mixed-mode switches can degrade performance drastically. Any frequently called parts of the OS that incur mixed-mode switches can cancel out the added performance provided by the Power Macs.

Native software can have it particularly hard if it calls emulated traps frequently. Each time an emulated trap is called from native software, a mixed-mode switch happens going into the routine and again coming back out. In System 7.1.2, the first release of the Mac OS for the Power Macs, some traps that you would have thought would be native aren't. For example, the Microseconds trap, used for fine-precision timing, is emulated. Calling this trap repeatedly from a native app yields not only unexpectedly long times, but also slows down the native software. Apple's steadfast refusal to release information about which traps are native and which are emulated seems a bit odd considering this example.

Apple's rationale is that people shouldn't count on whether a trap is emulated or native, since emulated traps are liable to go native without any warning. This is an understandable stance. But since one of the main goals of going native is increased software performance, and knowledge of which traps are emulated would allow developers to avoid those traps that would slow them down—possibly in favor of a solution that is already native—it's not clear that this strategy accomplishes anything useful. Regardless of Apple's position on this matter, at this writing, a list of native, fat, and emulated traps had already been posted by an enterprising developer to numerous online services.

Call Chains

A mixed-mode switch rarely comes alone. For this reason, determining the effect of making a particular piece of software native or leaving it in emulation on a Power Mac requires careful investigation to determine how the code is called, and where execution continues after the code in question is completed. A *call chain* is the path of execution taken when a call to a particular system-software routine is made. Many different bits of code are executed after a piece of software calls a system-software routine and before the code of the routine itself is executed. Each separate routine of intervening code makes up a link in the call chain. To make matters even more complicated, not only are these different intervening routines executed going down the chain from caller to called routine, but different parts of those same routines are generally executed on the way up the chain, back to the original caller, after the system-software routine has done its job.

Since each link in the call chain can consist of either PowerPC or 68k code, you may incur a mixed-mode switch once for every routine in the chain. On the way back up, the same number of mixed-mode switches happens again.

When third-party software—or even Apple software—patches into such a call chain, any additional mode switches that result from the patch can bog down the entire machine. Some operating-system routines are so popular that they are sometimes patched multiple times, each time by a different extension or control panel. In the worst case, the different patches in a single trap use both PowerPC and 68k code, causing many extra mixed-mode switches.

If you're a developer who patches traps, you owe it to your users to investigate this issue in great detail and determine the best method for minimizing the impact of any mixed-mode switches. If you're a user, you should find out whether any of the extensions that you use regularly cause unnecessary mixed-mode switches by patching emulated code into a native trap.

Extensions and Fat Patches

Extensions and control panels are an unavoidable part of today's Macintosh experience, but they aren't inherently evil. What gives these pieces of software such a bad reputation is programmers' often shoddy programming practices that manifest themselves as INIT conflicts. In general, it is a good idea to use as few third-party extensions and control panels as possible. Even on a 68k-based Macintosh, extensions and control panels cause a slowdown just because of the added work that they perform.

On the Power Macs, the incentive to avoid unnecessary extensions is even greater, since emulated software runs slower than native software anyway, and the likelihood is high that extensions are causing mixed-mode switches and thereby degrading the overall performance of a Power Mac.

During the migration from 68k to PowerPC, which will undoubtedly take many years, extension authors need to be aware of the choices they face about how to patch system software.

Relying on the fact that a trap is currently emulated isn't necessarily a wise thing, since it can become a native trap without warning as soon as the next system software release, or even a new version of the PowerPC Enabler, comes out. The "but the trap's emulated" excuse will hold water for a while, but not for much longer.

Native traps—any of the QuickDraw calls, for example—should always be patched native. There is no excuse for slowing down users' machines by introducing 68k QuickDraw patches on a Power Mac system. QuickDraw accelerator cards are a good example of this. By now, all the major video card vendors have added Power Mac support to the drivers on their video cards. Those accelerators that don't have Power Mac support wind up patching Native QuickDraw with the intent of speeding it up, but by adding a 68k patch to a native QuickDraw call, the accelerator's software is actually slowing everything down. If you have a NuBus video card that you want to use in your Power Macs, make sure the card has the latest ROMs from the vendor and that any additional Power Mac–specific drivers are installed. This will minimize any chance of slowdown related to mixed-mode switches.

In general, extensions these days should install *fat patches*. These patches contain both 68k and PowerPC code and have the benefit of never causing an unnecessary mixed-mode switch. When an extension installs a fat patch and the patched trap is called, the Mixed Mode Manager looks at the patch to see what kind of a patch it is.

- If it's an emulated patch and the caller is native, a mixed-mode switch must happen.
- If it's a fat patch, the Mixed Mode Manager picks the code type of the caller. If the caller is emulated, the emulated patch code is executed. If the caller is native, the native patch is run.

The only time when it makes sense to patch 68k traps exclusively with a native patch is if more time is saved executing the patch than it takes for two mixed-mode switches to occur. If this is the case, then a native patch is a fine idea.

It's safe to assume that system-software calls that are native today will remain so. Therefore, if you have to patch such a trap, make sure your patch is native; a fat patch won't be much help. Even if the rest of your code is still emulated, make sure that the piece of code that determines whether the rest of your code should be executed is native. This is the strategy Farallon's engineers used starting with version 1.0.3 of Timbuktu Pro.

Timbuktu Pro intercepts QuickDraw calls and retransmits them over the network to another Mac, allowing the user of the remote Mac to see what's happening on the local Mac's screen. To intercept the QuickDraw routines, Timbuktu must patch them. Since QuickDraw is native on the Power Macs, using 68k patches would slow the machine down measurably. Farallon's solution was to install native patches that didn't incur mixed-mode switches. Only when a remote Macintosh is connected to the local Mac does Timbuktu need to intercept and retransmit the QuickDraw information.

Since Timbuktu Pro 1.0.3 isn't otherwise native, mixed-mode switches happen when someone is connected remotely; the code to capture the QuickDraw information and send it is still emulated. But if no one's watching, Timbuktu's patches stay dormant and do nothing. As a result, the patches cause no mixed-mode switches. This is the ideal strategy for this kind of situation. Ideally, the rest of Timbuktu would be native as well, but the need for a native version has been alleviated since Timbuktu's patches no longer affect the Power Macs' overall performance.

The Code Fragment Manager

The Code Fragment Manager (CFM) is a crucial new piece of system-software technology introduced with the Power Macs. The initial implementation of the CFM is focused on the PowerPC, but a 68k version of the CFM will become available before the end of 1994.

When the Macintosh operating system was originally designed, RAM was a scarce commodity, and virtual memory required far more computing resources than were available. The designers of the original Mac OS came up with a scheme that would allow applications to load only the code they really need into RAM and leave unused code on the disk to be retrieved later if needed. In this scheme, known as *segmentation,* each individual piece of code is referred to as a *segment*.

Over the years, the Mac operating system's segmentation scheme has become a hindrance to many developers. The state of the Mac has also progressed. From experience gained in the intervening years, a definite need has arisen for a new piece of system software that is responsible for the same things the segment loader was, but that is far cleverer and more modern about it. Enter the Code Fragment Manager.

Rather than storing executable code in small 32-kilobyte chunks, native PowerPC applications store their executable code in a contiguous chunk in the data fork of the application file. This new scheme also allows fat binary applications—apps that contain both 68k and PowerPC code and that can run on any kind of Macintosh at the best possible speed.

Storing the PowerPC code in a single piece allows the implementation of a useful feature on the Power Macs: code swapping. When virtual memory is enabled on a Power Mac, any native app that is launched has only the code that it actually needs loaded into RAM. Code needed later is loaded into memory with the help of the Virtual Memory Manager, which treats the data fork of the native application as if it were a mini VM swap file. With this scheme, PowerPC-executable code is also made to be read-only—the first time memory protection of any kind is available in the Mac OS.

Some apps store PowerPC code in external plug-in files. As long as the plug-in's code is in the file's data fork and the application software calls the appropriate parts of the CFM, the code-swapping feature is also available for nonapplication code.

Since the code will never be modified while it's running, anytime new code needs to be loaded into RAM, the operating system doesn't have to save the least recently used piece of code to disk; it just loads the new code. This makes the performance hit for using this scheme small, since reads from hard disks are always much faster than writes.

Another important feature of the Code Fragment Manager is that every fragment has its own global variables. This feature, provided for all types of PowerPC code that use the CFM, makes development of stand-alone code vastly easier. An application usually consists of a single fragment, but it can also be made up of multiple small fragments.

Under the standard 68k environment, stand-alone pieces of code had to go through all sorts of contortions to create a scheme for accessing global variables.

The final key feature of the CFM is support for shared libraries, known as *import libraries*. These libraries are not to be mistaken for shared libraries used by Apple's Shared Library Manager—they are two different things. The CFM has the ability to bind multiple fragments at runtime and allow one fragment access to code and data that has been explicitly exported by other fragments. CFM's support for import libraries lets you keep a single copy of core code around that is shared by multiple applications in the same family. But that's not all. CFM's import libraries can also act as update libraries, providing replacements for existing code. This would allow an application developer to provide updates simply by making update libraries available rather than sending an entirely new app. With the update library scheme, only code that needs to be overridden is provided, and the CFM deals with all the trickiness involved in reconciling different version of libraries and making sure that the most recent version of a routine is called.

The CFM for PowerPC is a welcome addition to system software and provides extremely useful features to developers of native PowerPC software.

The Nanokernel

The nanokernel in the first PowerPC version of Macintosh system software is the lowest-level piece of system code and handles many of the hardware-specific tasks. It provides a layer of insulation between the hardware and the system software, allowing system software to use a standardized way of accessing certain low-level hardware features. When the hardware changes, the nanokernel must change as well to support the hardware, but the system software that calls the nanokernel should not have to change much, if at all. The nanokernel is a predecessor of the long-awaited microkernel for the Macintosh operating system.

Traps

On the 68k Macs, calling the operating system involves the use of traps. Traps are a method of interrupting software that's currently running. Trying to execute a particular type of instruction—for example, one whose hexadecimal form begins with the value A—causes an A-trap exception, which is dealt with by the A-trap handler. The A-trap handler on a Mac is known as the *trap dispatcher;* it looks at the value of the other 12 bits in the 16-bit instruction that begins with A and, based on the value, jumps to the part of the operating system that's being called by the 68k software. This routing of execution to the correct part of the OS is the trap dispatcher's job.

For native code, operating-system calls are dispatched via a different scheme that uses features of the Code Fragment Manager, but there is still a basic similarity. Traps can still be patched on the Power Macs, and the patches can be of three different varieties: a 68k patch, a PowerPC patch, or a fat patch.

There are four different types of traps on the Power Macs, described by the type of code that the OS dispatcher routes a call to.

Native Trap

Native traps are something of a misnomer, since a 68k trap is a native trap on 68k-based Macs. The dispatcher routes an OS call to a routine called a *native trap* that consists of PowerPC code. If a native trap is called by emulated software, it causes a mixed-mode switch before executing the trap's code, and another switch afterward before continuing execution of the emulated code.

Emulated Trap

The dispatcher routes the call to emulated 68k code. If the *emulated trap* is called by native software, it causes two mixed-mode switches: one before the trap's code is executed, and another afterward.

Fat Trap

The trap dispatcher has a table of addresses where it looks up where to route a call to. A *fat trap* has both PowerPC and 68k code; the decision about which type of code to use depends on the caller. Since mixed-mode switches exact a big performance penalty, the Mixed Mode Manager endeavors to avoid causing a switch where possible. For this reason, if emulated 68k code calls a fat trap, the code executed is also emulated 68k code, thus avoiding two mixed-mode switches. If PowerPC code calls a fat trap, PowerPC code is executed, again to avoid the switches.

Fat traps are the ideal solution during the transition from 68k Macintosh to Power Macintosh; a fat trap causes the fewest mixed-mode switches.

Split Trap

A *split trap* denotes an OS routine whose native PowerPC version doesn't go through the central OS dispatcher. For this reason, there is no way for third-party software to patch such a trap. Even if the patch is installed, it has no effect on the execution of the native OS call. Split traps generally occur where Apple engineers felt that there was no good reason for any software to patch that routine, and avoiding the overhead of the dispatcher provides a small performance boost as well.

The PowerPC system software contains significant changes when compared with Macintosh system software running on 68k. The Mixed Mode Manager transparently takes care of much of the work involved in calling PowerPC code from 68k code and vice versa. The Code Fragment Manager provides a new runtime environment for PowerPC native apps and stand-alone PowerPC code. The CFM in particular is the first sign of some major changes to the Macintosh operating system to come.

Floating-Point Writes Go Fast

On a Power Mac, Native QuickDraw uses an interesting way to double its read and write performance when moving large amounts of memory around, for example when transferring a block of image data from one part of the screen to another. The trick is in putting together two details: the PowerPC 601's data bus is 64 bits wide, and so are floating-point registers (FPRs).

Experimentation has shown that a significant performance increase can be seen when using floating-point writes to memory that is non-cacheable, such as video memory, rather than doing the same thing from 32-bit GPRs (general-purpose registers). The trick is this: Data is written from two GPRs into a location in memory that is actually cached in writeback mode. These two 32-bit values are read back into a 64-bit FPR and then written out again in a single beat, this time to the location in memory where the data is supposed to go.

This method won't show a clear benefit for moving small amounts of data around on a Power Mac, and it's detrimental when writing to cacheable memory, but if your software spends much of its time writing large amounts of data to noncacheable memory, you should definitely look into those big floating-point registers.

CHAPTER NINE

Looking Ahead

he first generation of Power Macs hit the mark by providing a great performance enhancement over 68k-based Macs when running native applications, all the while maintaining hardware compatibility with existing Macintosh peripherals. However, overall system performance is determined by more than just the performance of the system's central processor. Despite the high-performance design of the Power Macs' 64-bit CPU bus and DMA hardware, some carryovers from the 68k Mac that were already known as serious bottlenecks have also come along for the ride, and they remain sore points.

Apple has already announced its intention to support forthcoming technology standards such as PCI and FireWire, both of which offer features and performance that is much more in line with the increased speed offered by native PowerPC applications. In addition to performance boosts in the midrange and high end, PowerPC processors will migrate across Apple's entire product line.

The current Power Macs are the starting point for many major developments to come, within the Macintosh industry as well as outside of it. This chapter takes a look at some of the major new hardware and software technologies that are on the way and their impact on the Macintosh and the rest of the personal-computer industry.

Hardware

Processors like the 100MHz 601 and the 604 run software very quickly on their own, but the applications that benefit the most from these PowerPC chips' computational horsepower are also dependent on I/O performance. Multimedia applications such as digital video and audio are a major burden on the video and storage subsystems in a Mac, and anything that improves performance in these areas is a major boon.

PCI and FireWire are two technologies that Apple has openly committed to supporting in the Macintosh line—perhaps as soon as in the next generation of the Power Macs. Not only are these technologies solutions to existing bottlenecks in the standard Macintosh hardware, but they are also multiplatform standards. The Mac won't be the only personal-computer system using peripherals based on these schemes (unlike SCSI, which, until the popularity of CD-ROM in the x86 world, was virtually unheard of there). With multiplatform support comes larger production and sales volumes for these products and, consequently, lower prices and a broader selection of products for Macintosh users to choose from.

PCI

PCI (Peripheral Component Interconnect) is an expansion-card standard, analogous to NuBus on the Mac, that was originally developed by Intel but whose management has since been taken over by a vendor-independent organization called the PCI Special Interest Group. As the central contact point for vendors wishing to create PCI products, the PCI SIG has all the necessary information about technology licensing and related issues.

PCI is being evangelized as an open standard. Early fears that Intel would rule over PCI heavy-handedly and require exorbitant licensing fees and royalties have proven unfounded. In fact, Digital, maker of the Alpha family of RISC processors, announced an Alpha chip in late 1993 that contained PCI interface logic directly on the chip. This bodes well for future microprocessors of other families that might wish to integrate PCI directly on-chip as well, thereby reducing system cost by obviating the need to use PCI interface chips in the system design.

Basic Features: PCI is an expansion bus designed to allow peripheral cards to be added to a computer system. Today's Macs' NuBus will be replaced by PCI in future Power Macs. It is possible to use PCI as a bus on a system's motherboard, but common PCI use today is for expansion cards. Future hardware designs might access a motherboard video subsystem via the PCI expansion bus.

Like NuBus, but unlike the common expansion buses in the x86 world, PCI cards are self-configuring. They require no setting of dip switches or jumpers.

The initial version of PCI has a 32-bit-wide bus and runs at 33MHz, providing maximum theoretical throughput of roughly 126MB per second. NuBus' theoretical maximum is approximately 38MB per second, or twice that when moving data between two cards on the same NuBus.

NuBus has long been a major bottleneck for video cards on the Mac. PCI's added bandwidth should remove that problem and allow new kinds of video cards that aren't possible with NuBus' limited bandwidth.

The current PCI specification already defines a 64-bit version of the bus, doubling the theoretical maximum throughput to approximately 252MB per second—plenty even for the most bandwidth-hungry applications.

PCI cards can use one of two voltages for power: 5V or 3.3V. See Figure 9.1. The current standard for desktop computers is 5V, but 3.3V is rapidly gaining popularity because

FIGURE 9.1
PCI with 3.3V and 5V connections

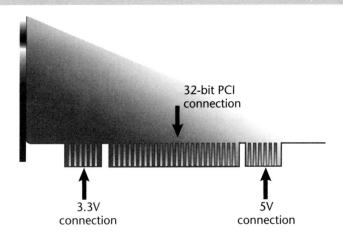

it is the standard voltage used in laptops and mobile computers. Also, 3.3V is gaining increasing support in the desktop-computer world because of the popularity of power-conserving "Green" PCs.

The mechanical specification for PCI cards ensures that the wrong card can't be plugged into a PCI slot. However, support for the two voltages is designed in such a way that it doesn't make them mutually exclusive: It's possible for a single PCI card to support both 5V and 3.3V power. Some cards already have this support.

Drivers: PCI promises to allow a single card to operate in many different hardware and system software environments, since nothing about the PCI hardware ties it to any particular microprocessor architecture. However, the issue of PCI drivers is a thorny one.

A Power Macintosh with a PCI slot is a very different operating environment than a Windows-based Pentium-PC with a PCI slot. The hardware is identical, but the operating systems have completely different I/O architectures that aren't even slightly compatible. The purpose of a PCI driver is to allow operating systems to use their standard APIs to access the cards' features. But given multiple operating systems and hardware environments, how can the card know which driver is the right one?

The answer is to store the PCI driver on a system's hard

Open Firmware

Open Firmware is the colloquial name for the IEEE standard number 1275–1994 for boot firmware. Open Firmware's ancestor OpenBoot was originally developed at Sun Microsystems in 1988 when Sun was shipping machines based on three different processor architectures. Sun needed a standard method for booting its systems that would work equally well on all of its systems. Version 1 of Sun's OpenBoot software was introduced with Sun's SPARCstation 1 workstations. Version 2 of OpenBoot, which is the version that the draft standard of Open Firmware was based upon, was first introduced with Sun's SPARCstation 2 machines.

Open Firmware is designed to provide an operating-system- and processor-independent method for booting a computer system. During the boot process, peripheral devices must be identified and their drivers loaded, and when all the hardware is initialized, the operating system must be loaded and started. Once the operating system is launched, Open Firmware has completed its work.

Since Open Firmware must be processor independent, its native language is interpreted and based on the programming language Forth. Open Firmware drivers written in FCode, as Open Firmware's Forth derivative is known, can operate in any Open Firmware environment, since every Open Firmware implementation contains the FCode interpreter.

When an Open Firmware-based system boots and identifies the devices connected to the system, it builds a *device tree*. This data structure contains entries for all devices that Open Firmware has identified. An operating system can later traverse the device tree to determine the available hardware.

Open Firmware and PCI form a symbiotic relationship and allow PCI's multi-platform driver problem to be solved. Since PCI is a platform-independent standard, there is no way of knowing in advance what kind of processor is available in the system that the PCI card is plugged into. Putting an FCode driver in the PCI card's ROM allows any Open Firmware-based system to initialize the PCI card and use it.

disk and have it load at boot time, instead of loading the driver from the card's ROM. It is certainly possible to store a card's driver(s) in ROM on the card itself, but in the future, including one driver for every PCI-capable hardware and operating-system configuration will be unworkable. The amount of ROM required for this would raise the card's price unnecessarily. Storing the driver on a local hard drive guarantees that the correct driver is loaded for the card, and it also allows easy upgrading of a card's drivers—much easier and less expensive than replacing a card's ROM. In this scenario, the card's ROM must contain only the necessary information to allow the boot firmware of the system that the card is installed in to identify the card so that the correct driver can be loaded. The Open Firmware standard, a platform-agnostic scheme for booting a computer system and configuring its peripherals at boot time, has provisions to support this method of driver loading at boot time.

PCI will bring high-speed peripherals back into line with the additional performance offered by the Power Macs' PowerPC processors. For Mac users, the switch to PCI is an all-around win: installation will be as hassle-free as with NuBus, performance will be higher, and card prices are likely to be lower, since PCI card manufacturers can build one card for all PCI markets and need only provide driver software for the Mac.

FireWire: FireWire is another new high-speed I/O technology that Apple has publicly committed to support. In the long run, FireWire may replace today's SCSI for access to external mass-storage connection.

FireWire is the Apple-trademarked name for the IEEE draft standard P1394. FireWire's goals are to provide a low-cost, high-performance, plug-and-play peripheral bus to connect a computer system and external high-speed peripherals. See Figure 9.2.

FIGURE 9.2
FireWire

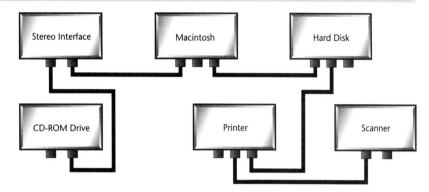

Compared to SCSI, FireWire has significant advantages:

- Allows hot connections: You don't need to power down all devices on the bus to add one.
- Fast: Current implementations provide throughput of 98.3 megabits (roughly 12.3MB) per second. This is faster than today's Fast SCSI implementations.
- Small: The connectors and cables are tiny when compared to SCSI cables.
- Real-time: P1394 supports isochronous data transfers. This means that time-critical data, such as a QuickTime movie or digital video content, plays back over FireWire with no drop-outs.
- Multiple masters: FireWire devices can communicate and transfer data between themselves without the computer system being the midpoint in the transaction. Data is transferred from point to point between the devices.

FireWire has some additional features that those familiar with the trials and tribulations of functioning SCSI buses will love.

- Topology: FireWire doesn't need to be a strict chain as with SCSI. Any sequence of connections is fine, as long as the entire FireWire chain doesn't form a closed loop.

- Termination: No explicit termination is necessary; it's handled automatically.
- No ID conflicts: FireWire devices identify themselves on the bus and arbitrate a free ID without user intervention.

With these many features, FireWire appears to be the perfect external high-speed peripheral bus. The only current catch with FireWire is that support for it is beginning slowly. Manufacturers of mass-storage devices need to integrate FireWire interfaces into their controllers. Several companies have announced the availability of FireWire chip sets for this kind of application, but most vendors are waiting to see whether FireWire takes off before investing time and resources into this new technology. FireWire will remain dormant until some major computer and expansion-card manufacturers ship FireWire interfaces for popular computer systems. Without support for it on the system side, there's no sense in making peripherals for it.

FireWire is a clearly superior technology and should have no problem supplanting SCSI in the long run. In the short term, however, it will be difficult to convince systems and peripherals vendors to support the new technology and produce sufficiently inexpensive solutions to get the migration started.

The PowerPC Reference Platform

The PowerPC Reference Platform is a hardware and system software specification developed by IBM and Motorola. Its purpose is to provide guidelines for the implementation of PowerPC-based personal computers, so that PowerPC-based systems from multiple vendors remain as compatible as possible with each other.

Unlike many specifications, the PowerPC Reference Platform document doesn't go into the nitty-gritty implementation details of each of its features. The intent with the specification is to provide a set of features for the lowest-common-denominator PowerPC-based system. Exactly how the features are implemented is left up to the individual manufacturer.

The PowerPC Reference Platform specification also includes details about a system-abstraction layer (SAL) of firmware that provides a standard API for operating systems to access Reference Platform hardware features. The idea is to be able to buy a shrink-wrapped Reference Platform operating system at the superstore of your choice and be able to install it successfully on any Reference Platform-compliant machine.

The first batch of Power Macs does not comply with the Reference Platform. Their design commenced well before the Reference Platform effort got under way. Whether future Power Macs will comply with this standard is unclear. The specification requires every machine to have a parallel port, something for which there is no need at all in the Macintosh universe, and at this writing, the Reference Platform does not include support for the Apple Desktop Bus, which is used to connect keyboards, mice, and other input devices.

There is in many quarters the hope that Apple will produce a Reference Platform version of the Macintosh system software, to broaden the Macintosh market significantly. Such a move would make a great deal of sense for the Macintosh market as a whole. And even if there were PowerPC clones capable of running the Mac OS, Apple's hardware would in all likelihood still be the best for running the Mac OS, because of the close integration of Macintosh system software with new hardware features such as the AV capabilities.

No major PC vendors other than IBM had jumped onto the PowerPC Reference Platform bandwagon by mid-April 1994. Numerous operating systems for the Reference

Platform were already being planned to support the standard PowerPC hardware, however, among them Microsoft's Window NT, IBM's Workplace OS, and SunSoft's Solaris.

The long-term success of the Reference Platform is unclear. The idea of a standard PowerPC hardware specification is a good one, but good ideas alone don't guarantee success. The basic OS support is there: IBM has a bridge/migration strategy for existing x86-based Windows users, and the rest is up to the buying populace.

Hardware

The initial version of the Reference Platform specifies a 6xx series central processor. It has provisions for all the standard I/O ports such as SCSI, Ethernet, a parallel port, and even LocalTalk. There is no standard expansion bus, although it looks like PCI will be the de facto standard. The Reference Platform specification is so flexible that the aging ISA PC bus is supported for low-speed peripheral cards. In general, the Reference Platform prototype designs show philosophical similarities with current x86 PCs. Since x86 users are the target market for these machines, it makes sense to provide as familiar a transition environment as possible.

Software

The software side of the PowerPC Reference Platform is in many ways more interesting than the hardware. Each Reference Platform system comes with enough firmware to initialize the hardware and load a compliant operating system. The Reference Platform specifies the use of 1275–1994 Open Firmware (described in the "Open Firmware" sidebar) for the initial startup process. Low-level drivers for peripheral devices can be provided either as FCode or in a specific operating-system-dependent form.

Once Open Firmware has brought the hardware up and tested it, it hands over control to the chosen operating system. It will be possible to install multiple operating systems on a Reference Platform machine and choose between them at startup time. This flexibility and OS agnosticism will aid the adoption of the PowerPC Reference Platform hardware. Large corporations—the desired early adopters, mainly because they buy many machines at a time—are more likely to investigate new hardware if it offers a clear benefit over their existing systems. The anticipated ability to buy a standard PowerPC-based personal-computer system and install operating-system software as users require is appealing to large organizations that like to minimize the number of different systems they must support. For this reason alone, Reference Platform hardware warrants careful consideration by existing x86-based organizations.

Graphing Calculator

The Graphing Calculator that ships with every Macintosh is more than just a demo application to show off how quick the PowerPC's floating point is. It's a harbinger of the type of software that's in development now.

Although the calculator itself is useful enough to make it far more than a toy, its primary goal is to illustrate how user interfaces can benefit from the additional processing power afforded by the PowerPC. When you create a 3-D graph in the calculator, you can pick it up with the mouse and rotate the graph. On a 68k-based machine, an application like this would probably offer a wireframe rendition of the graph for rotation, since it would be too computationally expensive to rotate and redraw the entire graph. Not only does the Graphing Calculator allow the user to freely rotate the full nonwireframe graph, but the calculator is also recalculating every point on the graph during the rotation. It's not just moving video data around.

When a user types an equation in for the first time, the Graphing Calculator's display is a simple white space without a hint of graphing capability. But when the user hits the graph button, the divider bar that separates the equation area from the graphing area moves up the calculator's window, gradually revealing the graph behind. The movement of the divider is smooth, and the graph behind it is drawn as the divider goes up. There's no quick redraw at the end; the calculator is drawing the graph bit by bit, like an opening window shade that gradually reveals what's behind it.

Developers take note: Power Macintosh isn't just about software that crunches numbers faster. The available computational horsepower should be used to make Power Macintosh software even more user-friendly, more responsive, and more interactive. Above all, now that it's available, the Power Macs' performance should be used. This doesn't mean that developers should be wasteful with the computational power available. Applications such as the Graphing Calculator are examples of software that appears simple to the user, but a lot of thought, effort, and engineering have gone into making it so.

System Software

Early in 1994, Apple outlined milestones for the future of Macintosh system software: System 7.5, Copland and Gershwin, and OpenDoc. Power Mac support is a given for all of them.

System 7.5

System 7.5 is the first so-called reference release of Mac system software since the introduction of System 7.1 and will be available in summer 1994. System 7.5 combines several previously separate system-software products. It is the first reference release that contains support for the Power Mac. In fact, System 7.5 does not require system Enablers for any Macintosh CPUs that were shipped prior to the 7.5 release date. All the Enablers' functionality is now built in.

System 7.5 contains new pieces of system software, including Apple's long-awaited QuickDraw GX software, which offers new imaging, type, and printing capabilities. QuickDraw GX runs native on Power Macs. System 7.5 also comes with Apple Guide, the first part of Apple's strategy to offer a more active help system on the Macintosh. System software as well as applications can now provide all their documentation in an electronic format that also includes built-in tutorials that show the user exactly how to perform specific tasks. Software developers have to provide the necessary infrastructure to use Apple Guide to its fullest, but the user, especially those learning new software, will benefit tremendously from the software's ability to guide users through tasks.

System 7.5 integrates a large amount of system software technology that has been released in bits and pieces since the introduction of System 7.1:

- AppleScript
- A new scriptable Finder that can be controlled with AppleScript
- Macintosh Drag & Drop, which allows users to drag data between applications; for example, users can drag files from the Finder directly into open windows of some Drag & Drop–aware applications, obviating the need to use the Open item in the File menu
- Threads Manager, which offers software developers the ability to have multiple threads executing within their software
- QuickTime and the QuickTime PowerPlug

- Macintosh PC Exchange, which allows PC floppies and other removable media to be mounted in the Finder just like Mac disks
- MacTCP, the Macintosh implementation of the TCP/IP protocol stack; users at organizations that use MacTCP no longer require separate software licenses
- PowerTalk, the Macintosh implementation of the Apple Open Collaborative Environment; there will be no separate System 7.5 Pro just for PowerTalk

System 7.5 adds many small improvements, such as support for volumes larger than 2 gigabytes, in addition to the major enhancements listed here. Much of 7.5 was integration of existing technology, but the new features will both be helpful to Mac users and provide software developers with more opportunities to build easier-to-use products.

Copland and Gershwin

Copland and Gershwin are the code names for the next major releases of Macintosh system software. The ultimate goal is to move the Mac OS to a completely microkernel-based system that provides, among other things, memory protection between processes, preemptive multitasking, and high-performance I/O. As the Mac OS evolves away from dependence on the 68k architecture, more of the Mac system software will become native, boosting performance for Power Mac users via new system software.

Further details about these operating systems were not available at the time this book went to press.

OpenDoc

OpenDoc is a fundamental part of Apple's future software strategy, and even though it isn't dependent on the Power Macs, it will run on them and pave the way for a fundamental shift in software and use of personal computers.

In today's typical operating-system environments, documents are associated with specific applications, and multiple applications don't necessarily allow seamless integration of their different types of data in the same document. OpenDoc creates an environment where the document is the focal point for the user. A document is a virtual blank slate that can contain many different types of data. Each separate type of data is known as a *part* in OpenDoc parlance.

In the OpenDoc world, large monolithic applications are a thing of the past. Instead, there are *part editors,* one for each type of part. In an OpenDoc environment, a user can create a word-processing part and embed a spreadsheet or graphic part within it, or have the two follow one another.

OpenDoc is an open architecture managed by Component Integration Labs (CIL), an independent organization whose sole purpose is to support the proliferation of OpenDoc. CIL will act not only as the central repository of OpenDoc knowledge, but also provide validation and certification services to guarantee that OpenDoc parts work together. OpenDoc will fail if parts from different vendors don't work together. The user will come to expect to be able to use any number of different part editors together in the same document.

OpenDoc will be available on the Macintosh, Windows, OS/2, and some UNIX platforms. It uses technology developed at Apple as well as a technology from IBM. OpenDoc will be multiplatform from the outset and won't be limited to Macs at all.

OpenDoc will change the economies of software publishing drastically, since smaller developers will once again be able to compete with the software giants on a part-by-part basis. Since users can pick and choose the parts that suit their needs the best, they can use one vendor's word-processing part editor with another's spelling checker and have them work together seamlessly.

> Open Transport is the name for Apple's new network software architecture for the Macintosh that will be released in fall 1994. Open Transport will be available for 68k Macs, and it will be the only native networking software for Power Macs. The Power Macs currently run all their networking software in emulation.
>
> Open Transport (OT) will initially ship from Apple with support for the Appletalk and TCP/IP network protocols. Novell has announced plans to provide IPX/SPX support for OT, but at this writing, no announcements about availability had been made.
>
> Open Transport solves a large number of problems both for the user and for the developer. The big win for Power Macintosh users is that Open Transport's protocol stacks will run native on the Power Macs and support new features such as multihoming, the ability to use multiple network interfaces in the same Macintosh. OT will allow PowerPC-based Macs to be high-performance network servers, since all the protocol processing is performed in the native code, and with the addition of multihoming, the network interface ceases to be the bottleneck. OT will also provide backward compatibility and act like the familiar AppleTalk and MacTCP that software uses today. Existing AppleTalk and MacTCP software will be able to run with OT and benefit from OT's added performance without any need to change the software itself.
>
> For developers, Open Transport provides a single API that is applicable to all available OT protocol stacks. Today, developing AppleTalk-based software is quite different from the development of MacTCP-based software. With the introduction of OT, development for any OT protocol stack will be virtually identical. Networking software can support multiple protocol stacks quickly without anywhere near the effort required to develop for multiple protocol stacks on the Mac today.

The Future

In the fast-moving computer industry, it's difficult to make predictions because things change so quickly. There are, however, some safe Power Macintosh-related assumptions that can be made. The 603, or a 603 variant will be used in PowerBooks and laptops from IBM's Power Personal Systems division as soon as 603s are available in volume and as soon as the hardware designs can be completed. The 603 will also find its way into desktop machines from Apple and IBM's PPS. The 603 has a good price/performance ratio, despite its lower performance compared to the 601.

Since the Power Macs aim to stay at the forefront of personal-computer performance, 604-based Macs will also be created as soon as 604s are available in sufficient quantities.

Apple has gambled on the PowerPC, and it looks like it made a good bet. The alliance still seems to be functioning well, much to the surprise of most outsiders who would never have thought that three companies like Apple, IBM, and Motorola could work together without mishap.

If native software ships quickly enough, both on Power Mac and on IBM's Power Personal Systems machines, it can begin to capture the large Intel market. Emulation is an excellent migration and bridge strategy, but in the long term it'll be the native apps that convert users.

The future looks bright for Macintosh. The hardware is fast, the operating system is being overhauled to come up to speed with hardware developments, and for the first time in several years, there is palpable excitement in the Macintosh market. The sense of adventure has returned, as well as the desire to compete on even terms with an adversary that outnumbers the Mac by nearly an order of magnitude. Intel realizes the threat that PowerPC poses to its leadership position, and its advertising campaigns speak of its concern far more eloquently than any written analysis. The next 18 to 24 months will determine whether the Macintosh survives against overwhelming odds, whether it remains a niche machine, whether it is completely overrun by Windows running on Pentium, or whether it succeeds in gaining significant marketshare based on technical superiority and lower price/performance ratios.

APPENDIX A

Resources

f you are interested in learning more about microprocessor and computer architecture, or you are looking for a good reference on the subject, I cannot recommend the following book highly enough:

Computer Architecture, A Quantitative Approach
by John L. Hennessy and David A. Patterson
Published by Morgan Kaufmann Publishers
ISBN 1-55860-069-8

IBM publishes a condensed version of PowerPC books I to III that also doesn't contain any nonpublic information. Anyone interested in knowing about the PowerPC architecture in greater detail will find this book useful:

PowerPC Architecture
Customer Reorder Number 52G7487
Available from IBM at 800/426-6477, or via fax at 512/823-9467

User manuals and technical summaries about the individual PowerPC microprocessors are available from IBM and Motorola. If you are outside of the United States, contact your local IBM or Motorola sales office. Within the United States, you can contact

Motorola Semiconductor Products Technical Responsiveness Center: 800/521–6274

IBM Microelectronics: 800–POWERPC (800/769–3772) or via fax at 800–POWERfax (800/769–3732)

Two new *Inside Macintosh* volumes are available if you're looking for more information about the system software available on the Power Macs:

Inside Macintosh—PowerPC System Software
ISBN 0–201–40727–2
Inside Macintosh—PowerPC Numerics
ISBN 0–201–40728–0

Index

54CF96 chip, 178
6809 emulators, 154–155
68LC040 emulator, 138–143
  caching, 141
  cycles, 140–141
  EIEIO instruction, 141
  floating-point calculations, 142–143
  how it works, 145–146
  insurance, 139
  MOVE16 instruction, 140
  operation, 140–141

A

A/UX, 81
ABS (Apple Business System) hardware, 55
Abstract PowerPC, 108–111
ACE (Advanced Computing Environment), 8–9
Address, 86
Adobe Type Manager (ATM), 74
AMIC (Apple memory-mapped I/O controller), 172, 174
Antidependency, 90
Apple Adjustable Keyboard and Jaguar project, 4
Apple Business Systems software, 77
Apple Computer
  collaboration with IBM, 11–12
  RISC (reduced instruction-set computer) and, 3–21
Apple Desktop Bus (ADB), managing, 177
Apple Remote Access, 72
Apple Workgroup Server (AWS) 6150, 55
Apple Workgroup Server (AWS) 8150, 55
Apple Workgroup Server (AWS) 9150, 56
Apple/IBM/Motorola alliance, 12–21
  601 chip design beginings, 19–21
  integrating Macintosh into enterprise networking systems, 13
  Kaleida, 13
  landmark decisions, 16–17
  PowerOpen, 13
  PowerPC, 12
  Somerset design facility, 17–19
  Taligent, 12
AppleShare, 77
AppleShare Pro, 78
AppleTalk, 76–77
  protocol stack, 72
Architecture, 2
Ariel II chip, 177
Asynchronous I/O, 65
AudioVision connector, 42
AudioVision monitor and Jaguar project, 4
AV card, 44–45
  digital audio video (DAV) connector, 45
  memory, 46
AWACs (audio waveform amplifier and converter) chip, 176

B

Bandwidth, 173
BART, 166, 168
Branch folding, 105
Branch instruction, 87–88
Branch-history table (BHT), 130
Branch-processing unit (BPU)
  601 chip, 114, 117–118
  603 chip, 125
  604 chip, 131
Burst transaction, 89
Bus, 88–89
  arbitration, 88
  burst transaction, 89
  contention, 43, 88
  PowerPC 601 chip, 113
  PowerPC 603 chip, 121
  PowerPC 604 chip, 113, 127–128
  split transactions, 113
  traffic, 88
Byte assembly, 170

C

Cache-coherency protocols, 101

Caches, 110
  68LC040 emulator, 141
  associativity, 98
  blocks, 98
  coherency, 100–101
  copyback mode, 99
  direct-mapped, 98
  hit, 100
  Level 1, 40, 101, 162
  Level 2, 37, 39–40, 101, 160–163
  lines, 98
  microprocessors, 97–101
  minimizing contention, 98–99
  miss, 100
  PowerPC 601 chip, 113
  PowerPC 603 chip, 121
  PowerPC 604 chip, 127–128
  principle of temporal locality, 97
  set associative, 98
  SoftWindows, 151
  valid bit, 100
  write-back mode, 99
  write-through, 99
Call chains, 188–189
CD-ROM drives, 48
CISC (complex instruction-set computer), 2, 95–97
  contrasting traits, 95
  Pentium chips, 2
CIVIC (Cyclone integrated video controller), 180
Code segments and segmentation, 192
Code Fragment Manager (CFM), 193
Cognac project, 5–7
Cold Fusion project, 24
Color lookup table (CLUT), 177, 180
Completion unit (CU), 603 chip, 126
Copyback mode cache, 99
Cub card, 9–10
CUDA chip, 177–178
Curio chip, 174, 176
Cycles, 86
  68LC040 emulator, 140–141
Cyclone project, 21

D
Data Path chips, 43, 170
  PowerPC 601 CPU bus, 169
Decode/dispatch unit (DDU)
Die, 90
digital audio video (DAV) connector, 45, 181–182
Digital-to-analog converter (DAC), 177, 180
Direct memory access (DMA), 36–37, 177
Direct-mapped caches, 98
DRAM (Dynamic RAM), 37–39, 170–171
  expansion, 38–39
Drivers and PCI (peripheral component interconnect), 202–204
Dynamic RAM (DRAM)
  video, 41

E
EIEIO instruction, 108
  68LC040 emulator, 141
Emulated traps, 195–196
Emulators, 137–158
  6809, 154–155
  68LC040, 138–143
  compatibility, 148
  I/O, 146–148
  performance, 143–144
  SoftWindows, 148–153
  Wabi, 153, 155–157
Ethernet, 51
EtherTalk, 76–77
Exceptions, 93–94
Express Modem software, 75
Extensions and fat patches, 189–192

F
Fast SCSI bus, 49, 51
Fast SCSI throughput, 47–48
Fat binary applications, 193
Fat patches, 66, 189–192
Fat trap, 196
FireWire, 204–206
Floating-point calculations, 7, 79–80
  68LC040 emulator, 142–143
  software overview, 78–80

Floating-point digital signal processor (DSP), 35
Floating-point performance and PowerPC 601 chip, 34–35
Floating-point unit (FPU)
  601 chip, 114, 118–119
  603 chip, 125
  604 chip, 129
FPRs (floating-point registers), 197

G
GeoPort, 52, 175–177
  hardware, 175
  Jaguar project, 4
  software, 72, 74–76, 175–177
GeoPort Telecom Adapter, 52
GPRs (general-purpose registers), 197
Graphing Calculator, 209–210

H
Hardware, 31–35
  ABS (Apple Business System), 55
  CD-ROM drives, 48
  direct memory access (DMA), 36–37
  FireWire, 204–206
  future, 200–206
  GeoPort, 52, 175
  memory, 37–41
  NuBus, 51–52
  overview, 32–33
  PCI (peripheral component interconnect), 200–204
  ports, 52–53
  Power Macintosh, 159–178
  Power Macintosh AV Card, 179–183
  Power Macintosh Upgrade Card, 32, 53–55, 178–179
  PowerPC 601 chip, 33–35
  PowerPC 603 chip, 35
  PowerPC 604 chip, 35
  PowerPC Reference Platform, 208
  sound, 53
  storage and SCSI, 46–49, 51
  system, 160–178
  video, 41–46

VRAM Expansion Card, 182–183
High-speed memory controller (HMC), 162–164
Hurricane project, 20

I
I/O (input/output)
  emulators, 146–148
  memory-mapped, 100
  noncacheable addresses, 100
  Software overview, 71–73
I/O bus, 171–178
  54CF96 chip, 178
  AMIC (Apple memory-mapped I/O controller), 172, 174
  Ariel II chip, 177
  AWACs (audio waveform amplifier and converter) chip, 176
  CUDA chip, 177–178
  Curio chip, 174, 176
  Squidlet chip, 178
  SWIM III chip, 176–177
IBM collaboration with Apple Computer, 11–12
IEEE draft standard P1394, 204
IEEE standard 1275-1994, 203
Import libraries, 194
Industrial design and Jaguar project, 4
INITs and patches, 73–74
Insignia Solutions and Microsoft agreement, 149–150
Instruction set architecture (ISA), 94
instructions, 87
Integer unit (IU)
  601 chip, 114, 118
  603 chip, 124–125
  604 chip, 129
  register renaming, 124–125
Interrupt handlers, 94
Interrupts, 93–94

J
Jaguar project, 4–5
  Apple Adjustable Keyboard, 4
  AudioVision monitor, 4

GeoPort, 4
Industrial design, 4
other chips investigated by, 8–9
Pink operating system, 5
PlainTalk speech recognition, 4
Quadra 660AV, 4
Quadra 800, 4
Quadra 840AV, 4

K
Kaleida, 13

L
Latency, 93
LAW (Power Macintosh 7100), 24
Level 1 cache, 101
Level 1 cache RAM, 39
Level 2 cache, 101
   PowerPC 601 CPU bus, 161–162
Level 2 cache RAM, 37, 39–40
   SRAM (static RAM), 39–40
Lines, 98
Load/store unit (LSU)
   603 chip, 125
   604 chip, 130
LocalTalk, 77

M
MacHack, 25
MachTen, 81
Macintosh Application Environment
   (MAE), 83
Macintosh Application Services (MAS),
   82
Macintosh Display Card 8•24 GC, 3
Macintosh IIfx, direct memory access
   (DMA), 36
MacTCP, 72
Memory
   AV card, 46
   Dynamic RAM (DRAM), 37–39
   hardware, 37–41
   Level 2 cache RAM, 37, 39–40
   page swaps, 70

PowerPC 601 chip, 113
PowerPC 603 chip, 121
PowerPC 604 chip, 127–128
ROM (read-only memory), 38, 41
software overview, 69–71
virtual memory (VM), 37–38, 40–41
VRAM (video RAM), 38
Memory addresses, 35
Memory-mapped I/O, 100
MESI protocol, 113–114
Mickey chip, 181
Microkernel operating system, 22
Microprocessors, 85–101
   address, 86
   architecture, 94–97
   basic concepts, 85–94
   branch instruction, 87–88
   bus, 88–89
   caches, 97–101
   CISC (complex instruction-set computer),
      2, 95–97
   cycles, 86
   dependencies, 90
   die, 90
   exceptions and interrupts, 93–94
   implementation, 97
   instruction set architecture (ISA), 94
   instructions, 87
   latency, 93
   pipelines, 90–93
   pointers, 86
   registers, 86–87
   RISC (reduced instruction-set computer),
      2, 95–97
   superscalar, 93
   transistors, 89
   wafers, 89
Mixed mode, 62–66
Mixed Mode Manager, 6–7, 21, 62, 185–188
   68k to PowerPC mode, 186
   fat patches, 66
   penalty for switching, 187–188
   PowerPC to 68k mode, 187
   routine descriptor, 186

Index 223

transition vector, 186
Mixed-mode switches, 62–66
   INITs and patches, 73
Modern Memory Manager, 69–70
   virtual memory (VM) and, 70
Motorola 88110 RISC chip, 7, 9–11
MOVE16 instruction, 140, 167–168
Multiply-add-fused (MAF) instruction, 106, 118, 125
Multiprocessing support, 113–114
MUNI (Macintosh Universal NuBus Interface), 166

N

Nanokernel, 194
Native PowerPC system software, 61–62
Native QuickDraw (NQD), 66–68
   doubling read and write performance, 197
   speed, 67–68
Native QuickTime, 68–69
Native traps, 195
NetWare, 78–79
Networking
   software, 76–78
   SoftWindows, 150
Noncacheable addresses, 100
NuBus
   bursting into the bus, 167–168
   card connections, 181–182
   excessive mixed-mode switches, 167
   hardware, 51–52
   performance, 167–168
   slots, 51
NuBus 90, 51, 168
   burst mode, 51

O

Open Firmware, 203
   PowerPC Reference Platform, 208
OpenBoot, 203
OpenDoc, 212–213
OpenTransport native protocol stacks, 78
Output dependency, 90

P

Page swaps, 70
PBBlockMove call, 168
PCI (peripheral component interconnect), 200–204
PDM project (Piltdown Man), 20
Pentium chips, 2
Pink operating system, 5
Pipelines, 90
   hazards, 91–93
   stages, 91
PlainTalk, 53
   Jaguar project, 4
Pointers, 86
Ports, 52–53
POWER architecture (performance optimized with enhanced RISC), 2, 11, 13–14, 103–106
   Power1, 104–106
Power Macintosh
   bandwidth, 173
   compatibility, 76
   Data Path chips, 43
   emulators, 137–158
   evolution and future, 28–29
   first demonstration of, 23
   future, 199–215
   hardware, 31–56, 159–178
   naming, 26–27
   performance, 56
   software, 57–83, 185–196
   system hardware, 160–178
   timelines, 27
Power Macintosh 6100, 31
   built-in video, 41–42
   CD-ROM drive, 49
   hard disks, 49
   HDI-45 AudioVision connector, 42
   internal bays for SCSI devices, 49
   NuBus adapter, 51–52
   SIMMs (single inline memory modules), 38
Power Macintosh 7100, 24, 31
   internal bays for SCSI devices, 49

processor direct slot (PDS), 44–45
SIMMs (single inline memory modules), 38
VRAM cards, 41, 46
Power Macintosh 8100, 24, 31
  Fast SCSI bus, 49, 51
  internal bays for SCSI devices, 49
  Level 2 cache RAM, 40
  processor direct slot (PDS), 44–45
  SIMMs (single inline memory modules), 38
  VRAM cards, 41, 46
Power Macintosh AV Card, 179–183
  CIVIC (Cyclone integrated video controller), 180
  DAV connector, 181–182
  Mickey chip, 181
  SAA7194 chip, 181
  Sebastian chip, 180–181
Power Macintosh NuBus Adapter card, 44
Power Macintosh Upgrade Card, 25, 32, 53–55, 178–179
  installing System 7.1.2, 58
  modes, 54
  performance, 54
Power1
  branch processor, 105–106
  chips, 104–105
  multiply-add-fused (MAF) instruction, 106
  vs. PowerPC, 111
PowerOpen specification, 13, 81–82
PowerPC, 12
  abstract, 108–111
  architecture, 109
  branch processor, 109
  caches, 110
  fixed-point execution unit, 109–110
  floating-point support, 110
  Somerset design facility, 17–19
  upgrades, 25
  vs. Power1, 111
PowerPC 403GA, 132–134

PowerPC 601 chip, 33–35, 111–120
  64-bit data bus, 35
  basic features, 113–114
  branch-processing unit (BPU), 114, 117–118
  bus, 113
  cache, 113
  Data Path chips, 43
  execution units, 114–119
  floating-point performance, 34–35
  floating-point unit (FPU), 114, 118–119
  high-speed cache, 35
  integer unit (IU), 114, 118
  memory, 37, 113
  memory addresses, 35
  multiprocessing support, 113–114
  speed, 33–35, 114
PowerPC 601 CPU bus, 160–169
  BART, 166–168
  Data Path chips, 169
  high-speed memory controller (HMC), 162–164
  Level 2 cache, 161–162
  processor direct slot (PDS), 165–166
  ROM (read-only memory), 164
PowerPC 603 chip, 35, 120–127
  basic features, 121
  branch processing unit (BPU), 124
  bus, 121
  cache, 121
  completion unit (CU), 126
  execution units, 124–126
  floating-point unit (FPU), 125
  integer unit (IU), 124–125
  load/store unit (LSU), 125
  memory, 121
  multiprocessing support, 121–122
  power management, 123
  sleaze mode, 125
  system-register unit (SRU), 125–126
PowerPC 604 chip, 35, 127–131
  basic features, 127–129
  branch-processing unit (BPU), 131
  bus, 113, 127–128

Index 225

cache, 127–128
decode/dispatch unit (DDU), 130–131
execution units, 129–131
floating-point units (FPU), 129
integer units (IU), 129
load/store unit (LSU), 130
memory, 127–128
multiprocessing support, 128
power management, 129
PowerPC 620, 134–135
PowerPC architecture, 2, 14, 217–218
PowerPC family, 103–135
   601 chip, 112–120
   603 chip, 120–127
   604 chip, 127–131
   abstract PowerPC, 109–111
   Books I-IV, 108–109
   instruction set, 107
   memory interaction, 107–108
   operating environment architecture, 109
   POWER architecture, 103–106
   PowerPC 403GA, 132–134
   PowerPC 620, 134–135
   virtual-environment architecture, 108
   what makes a PowerPC a PowerPC, 107–109
PowerPC Reference Platform, 206–209
   hardware, 208
   Open Firmware, 208
   software, 208–209
   system-abstraction layer (SAL), 207
Principle of temporal locality, 97
Processor direct slot (PDS)
   PowerPC 601 CPU bus, 165–166
   video, 44–45

Q

Quadra 660AV, 21
   direct memory access (DMA), 36
   Jaguar project, 4
   SCSI Manager 4.3, 65
   SCSI problems, 47
Quadra 800 and Jaguar project, 4
Quadra 840AV, 21
   direct memory access (DMA), 36
   Jaguar project, 4
   NuBus performance, 167
   SCSI Manager 4.3, 65
   SCSI problems, 47
QuickDraw, 22, 46
   native (NQD), 66–68, 197
QuickDraw GX, 211
   floating-point calculations, 34
QuickTime
   audio extraction features, 48
   native, 68–69
   version 2.0, 69
QuickTime PowerPlug, 68

R

RAM (random access memory) and cycle-stealing video, 43
Register files, 86–87
Registers, 86–87
RIOS. See POWER architecture, 13
RISC (reduced instruction-set computer), 2, 95–97
   Apple Computer and, 3–21
   Cognac project, 5–7
   common traits, 95
   Jaguar project, 4–5
   Macintosh Display Card 8.24 GC, 3
   Motorola 88110 chip, 7–11
   RLC (RISC LC), 5
   searching out chips, 7–12
   Sun Microsystems, 8
   system software, 21–22
RLC (RISC LC), 5, 10, 20
   Mixed Mode Manager, 6–7
   Standard Apple Numerics Environment (SANE), 7
ROM (read-only memory), 38, 41
   PowerPC 601 CPU bus, 164
Routine descriptor, 186
RSC (RIOS single-chip), 13, 106

S

SAA7194 chip, 181

SANE (Standard Apple Numerics Environment), 142–143
SCSI
  cabling, 47
  controller, 178
  driver software, 47
  handling port, 174
  internal bays for, 49
  termination, 47
  upgrading, 47
SCSI and storage, 46–49, 51
SCSI Manager 4.3, 47, 65
Sebastian chip, 180–181
Serial ports, handling, 174
Set associative cache, 98
SIMMs (single inline memory modules), 38–39
Smurf Card 19-20
Software, 57–83, 185–196
  Apple Business Systems software, 77
  call chains, 188–189
  Code Fragment Managager (CFM), 192–194
  compatibility, 76
  emulation, 60–61
  extensions and fat patches, 189–192
  fat patches, 66
  floating-point calculations, 78–80
  GeoPort software, 74–76
  I/O (input/output), 71–73
  INITs and patches, 73–74
  memory, 69–71
  mixed mode, 62–66
  Mixed Mode Manager, 6–7, 185–188
  Modern Memory Manager, 69–70
  native PowerPC system software, 61–62
  native QuickDraw (NQD), 66–68
  native QuickTime, 68–69
  networking software, 76–78
  PowerPC Reference Platform, 208–209
  System 7.1.2, 57–61
  traps, 195–196
  UNIX, 80–82
  virtual memory (VM), 70–71
SoftWindows, 148–153
  i486 emulation, 152–153
  cache, 151
  emulation strategies, 150–151
  I/O emulation, 151–152
  Insignia Solutions and Microsoft agreement, 149–150
  networking, 150
Somerset design facility, 17–19
  mixing corporate cultures, 19
Sound hardware, 53
Speed-bumping, 34
Split transactions, 113
Split trap, 196
Squidlet chip, 178
SRAM (static RAM), 39–40
Standard Apple Numerics Environment (SANE), 7, 79–80
  extended-precision format, 80
Storage and SCSI, 46–51
Sun Microsystems and RISC (reduced instruction-set computer), 8
Superpipelining, 91
Superscalar processors, 93
SWIM III chip, 176–177
SYNC instruction, 108
System 7.1.2, 21, 57
  basic features, 58
  installing, 58
  Mixed Mode Manager, 21
  toolbox acceleration, 59–60
System 7.5, 210–212
  QuickDraw GX, 211
System hardware, 160–178
  601 CPU bus, 160–169
  Data Path chips, 170
  DRAM, 170–171
  I/O bus, 171–178
System software
  Copland and Gershwin, 212
  diversification, 23–27
  future, 210–213
  microkernel operating system, 22
  OpenDoc, 212–213

RISC (reduced instruction-set computer), 21–22
System-abstraction layer (SAL), 207
System-register unit (SRU)
   603 chip, 125–126

T

Taligent, 5, 12
Tesseract project, 20
Timbuktu Pro, 191–192
Toolbox acceleration, 3, 59–60
Transistors, 89
Transition vector, 186
Trap dispatcher, 195
Traps, 195–196
   emulated, 195–196
   fat, 196
   native, 195
   Split, 196
True dependency, 90

U

UNIX, 80–82
   A/UX, 81
   MachTen, 81
   Macintosh Application Environment (MAE), 83
   PowerOpen specification, 81–82

V

V0 operating system and QuickDraw, 22
Video
   AV card, 41, 45–46
   built-in, 41–44
   cycle stealing, 43
   digital audio video (DAV) connector, 45
   DRAM (Dynamic RAM), 41
   hardware, 41–46
   processor direct slot (PDS), 44–45
   VRAM cards, 41, 46
Virtual memory (VM), 37–38, 40–41
   drawbacks, 70
   Modern Memory Manager and, 70
   software overview, 70–71

Virtual-environment architecture, 107
VRAM (video RAM), 38
   cards, 46
VRAM Expansion Card, 182–183

W

Wabi, 153–157
   version 1.0, 155–156
   version 2.0, 156–157
Wafers, 89
Write-after-read dependency, 90
Write-after-write dependency, 90
Write-back mode cache, 99
Write-through cache, 99